SpringerBriefs in Mathematics

SpringerBriefs in Mathematics showcases expositions in all areas of mathematics and applied mathematics. Manuscripts presenting new results or a single new result in a classical field, new field, or an emerging topic, applications, or bridges between new results and already published works, are encouraged. The series is intended for mathematicians and applied mathematicians.

More information about this series at http://www.springer.com/series/10030

Boštjan Brešar • Michael A. Henning
Sandi Klavžar • Douglas F. Rall

Domination Games
Played on Graphs

Springer

Boštjan Brešar
Faculty of Natural Sciences & Mathematics
University of Maribor
Maribor, Slovenia

Michael A. Henning
Department of Mathematics
and Applied Mathematics
University of Johannesburg
Auckland Park, South Africa

Sandi Klavžar
Faculty of Mathematics and Physics
University of Ljubljana
Ljubljana, Slovenia

Douglas F. Rall
Department of Mathematics
Furman University
Greenville, SC, USA

ISSN 2191-8198   ISSN 2191-8201   (electronic)
SpringerBriefs in Mathematics
ISBN 978-3-030-69086-1   ISBN 978-3-030-69087-8   (eBook)
https://doi.org/10.1007/978-3-030-69087-8

Mathematics Subject Classification: 05C57, 05C69, 91A43

This Springer imprint is published by the registered company Springer Nature Switzerland AG
The registered company address is: Gewerbestrasse 11, 6330 Cham, Switzerland

Preface

At its most basic level, this book is about dominating sets in graphs. A dominating set of a graph G is a subset D of vertices having the property that each vertex of G is in D or is adjacent to a vertex of D. Because of the many real-world problems that can be modeled by domination in graphs, research in the area has primarily focused on the problem of finding dominating sets of smallest cardinality. From a computational standpoint, this is known to be an intractable problem, and thus researchers have studied the problem when restricted to special classes of graphs or have produced upper and lower bounds on the minimum size of a dominating set in terms of other more easily computed graphical invariants.

When presented with a given graph and asked to find a small dominating set, it would seem natural to choose vertices one at a time based on some criterion such as large vertex degree. However, including vertices based on a single property (or even several properties) is not guaranteed to solve this optimization problem and in fact can lead to dominating sets that are far from being minimum. The domination game was introduced in 2010 as a way to model this vertex-by-vertex approach to building a dominating set of a graph. The game is played by two competitors who take turns adding a vertex to a set of chosen vertices. Each new vertex selected must enlarge the set of vertices dominated by the chosen set. The game ends when no such vertex exists in the graph, that is, once the chosen set is a dominating set. The two players have complementary goals, one seeks to minimize the size of the chosen set while the other player tries to make it as large as possible. The game is not one that is either won or lost. Instead, if both players employ an optimal strategy that is consistent with their goal, the cardinality of the chosen set is a graphical invariant, called the game domination number of the graph.

There are some surprises along the way. For example, there are connected graphs that have a spanning tree whose game domination number is much smaller than the game domination number of the original graph. This is perhaps counterintuitive since the domination number of a connected graph is always at most that of any of its spanning trees. Sometimes new methods and ideas that are developed within a theory are almost as important as its results. The first such method that we mention is the imagination strategy, which is the primary tool for proving results about domination game invariants. A perfect example of this is the proof of the Continuation Principle, a much used monotonicity property of the game domination number. In addition, we expose the discharging method of Bujtás. The power of this method was shown by improving the known upper bound, in terms of a graph's order, on the (ordinary) domination number of graphs with minimum degree between 5 and 50.

In the decade since it was published, the seminal paper [24] has 66 citations in MathSciNet. As a result, the game is quite mature with most of the notation and terminology commonly accepted. Hence we believe that now is the right time for a book on the game that summarizes the previous efforts, stimulates research on closely related topics, and presents a key reference for future developments. The book is intended primarily for research students in graph theory as well as established graph theorists, but it can be enjoyed by anyone with a modicum of mathematical maturity. Much has been done on game domination in the first 10 years, but many questions remain to be (asked and) answered, and a number of interesting conjectures are open. We hope this book can be a "jumping off" point for those interested in domination theory and in games played on graphs.

Prerequisites

A book of this size cannot be fully self-contained. Therefore, we assume some basic mathematical skills and do not explain fundamental concepts such as induction, sets, sequences, or functions. In addition, we assume a reader has some familiarity with basic concepts in graph theory such as connectivity and independence. Some knowledge of combinatorial algorithms would be useful, and in several situations, we discuss computational complexity, although these can be skipped without compromising the remainder of the topics.

Contents

Many of the proofs of results in the study of the domination games are technical and long. Because of space limitations, we are not able to include most of them. Thus, what we do instead is summarize what is known about various aspects of this research area. However, when possible, we do provide some of the proofs in order to give the reader an idea of the proof strategies involved when dealing with a game played on graphs. In all cases, we provide specific citations for the original papers. Some concepts covered in the book are more developed than others, but problems are suggested throughout.

In Chapter 1, we supply most of the graph theory definitions and notation used in the remainder of the book. The domination game and the total domination game are introduced and we meet the two players, Dominator and Staller. Together these two games define four graphical invariants for each graph depending on which player starts and which game is being played. For a small graph, we construct the entire game domination tree, and thereby illustrate why these numbers depend only on the graph. We also present natural bounds relating the game (total) domination number and the (total) domination number of a graph.

Chapter 2 is devoted to the standard domination game. Proofs in the realm of the domination game are often quite different from proofs of results in domination theory. Here we present the imagination strategy and illustrate its use in proving the Continuation Principle. Two of the outstanding conjectures, the $\frac{3}{5}$-conjectures, concerning upper bounds on the game domination number are presented, and we prove a weaker upper bound while illustrating the discharging method of Bujtás. Although their nontrivial proofs are not included, the formulas for the game domination numbers of paths and cycles are given. In this chapter, we also present what is known about the domination game played on subgraphs and trees as well as graphs whose game domination numbers attain the theoretical minimum or maximum values as a function of their domination number. Chapter 2 is concluded with a presentation of the computational complexity of the natural decision problems associated with the game domination numbers.

Chapter 3 is devoted to the total domination game. At this time, not as much is known about the total domination version of the game, but this chapter covers topics in a nearly parallel fashion to Chapter 2. We present the Total Continuation Principle and the best known upper bound for the game total domination number of a graph (all of whose components have at least three vertices) in terms of its order. The main conjecture in this area is the $\frac{3}{4}$-conjecture. The truth of this conjecture in some large classes of graphs has been verified, and the strongest of these is presented, without proof. Versions of critical and perfect graphs with respect to the game total domination number are introduced and known results are summarized. A complete characterization is given of trees whose game total domination number is equal to the total domination number. The chapter ends with a presentation of the computational complexity of the game total domination number.

In Chapter 4, we consider several new invariants that have arisen from a modification of the domination games. Given a linear order of the vertices of a graph G, an online (total) domination algorithm would select a vertex v precisely when at least one vertex in the (open) closed neighborhood of v is not (totally) dominated by the vertices preceding v in the order. The largest number of vertices selected by such an algorithm over all possible orderings of the vertex set is called the Grundy (total) domination number of G. Equivalently, this is the number of vertices chosen in the (total) domination game if both players adopt Staller's goal. Some upper and lower bounds for these numbers are given and the effects of vertex or edge deletion on these invariants are also studied. There are graphs for which every permutation of the vertices results in a (total) dominating set of the same cardinality. A characterization of some classes of these graphs is presented. Several other versions of these dominating sequences are introduced, one of which is related to the concept of zero-forcing in graphs, which in turn has connections to linear algebra.

Chapter 5 introduces and summarizes what is known about more than a dozen other games played on graphs or hypergraphs. Most of these are either modifications of the domination games presented in the first four chapters or are related games played on hypergraphs, which can in some instances lend insight into the domination games on graphs. An example of the former type is the connected domination game in which the set of vertices chosen by the two players at any point in the game must not only satisfy the rules of the domination game but must also induce a connected subgraph. The transversal game on hypergraphs is an example of the latter. In all of these games, there are problems waiting to be solved, so the area is rich for further research.

Acknowledgments

We are indebted to many colleagues and students who read parts of the book, gave useful remarks, or kept us informed about recent developments in the area and to those who provided technical support. In particular, we express our heartfelt thanks to Csilla Bujtás, Tanja Gologranc, and Vesna Iršič. Gašper Košmrlj is due a special acknowledgment for allowing us to use his software written specifically for analyzing these domination games. We also acknowledge the financial support from the Slovenian Research Agency (project J1-9109 and research core funding No. P1-0297). In addition, we thank Elizabeth Loew, Executive Editor at Springer, for her help and patience in guiding us through the process of preparing our manuscript for publication. Last, but not least, we all thank our families and friends for their understanding, patience, and support.

Maribor, Slovenia	Boštjan Brešar
Johannesburg, South Africa	Michael A. Henning
Ljubljana, Slovenia	Sandi Klavžar
Greenville, SC, USA	Douglas F. Rall

Contents

Chapter 1
Introduction

In this chapter, we first list the concepts that are essential for the book. These include domination and total domination, further graph invariants, graph products, and hypergraphs. After that we introduce the key players of the book, the domination game and the total domination game. The two games are not games in the sense of combinatorial game theory where one has winners and losers, instead they lead to the game domination number and the game total domination number. Using the game tree of a domination game we demonstrate that these numbers are well-defined graphical invariants. We conclude the introduction by presenting some basic bounds on both game invariants in terms of the (total) domination number.

1.1 Some Useful Notions and Notation

Let $G = (V(G), E(G))$ be a finite graph. The *order* and the *size* of G are denoted by $n(G)$ and $m(G)$, respectively. A graph G with $m(G) = 0$ is a *totally disconnected graph*, and a graph with $n(G) = 1$ is *trivial*. If $v \in V(G)$, then the *(open) neighborhood*, $N_G(v)$, of v in G is the set of vertices adjacent to v. The *degree*, $\deg(v)$, of v is $|N_G(v)|$. The minimum and the maximum degree of G are denoted by $\delta(G)$ and $\Delta(G)$, respectively. A vertex of degree 0 is an *isolated vertex*, while a vertex of degree 1 is a *leaf* or a *pendant vertex*. A graph is *isolate-free* if it has no isolated vertices. The unique neighbor of a leaf is called the *support vertex* of the leaf. A support vertex is a *strong support vertex* if at least two of its neighbors have degree 1. The *closed neighborhood* of v in G is defined by $N_G[v] = N_G(v) \cup \{v\}$. For a set $S \subseteq V(G)$, the *(open) neighborhood* of S in G is defined as $N_G(S) = \cup_{v \in S} N_G(v)$ and the *closed neighborhood* of S in G is defined as $N_G[S] = N_G(S) \cup S$. When there is no possible confusion we may omit the subscript G in the above notation.

© The Author(s), under exclusive license to Springer Nature Switzerland AG 2021
B. Brešar et al., *Domination Games Played on Graphs*, SpringerBriefs in
Mathematics, https://doi.org/10.1007/978-3-030-69087-8_1

If X and Y are subsets of vertices in a graph G, then the set X *dominates* the set Y in G if $Y \subseteq N[X]$. When X consists of a single vertex x we say that x *dominates* each vertex from Y, where clearly $Y \subseteq N[x]$. In particular, if a set X dominates $V(G)$, then X is a *dominating set* of G; that is, every vertex in $V(G) - X$ is adjacent to a vertex in X. In other words, X is a dominating set of G if its closed neighborhood is $V(G)$. The *domination number*, $\gamma(G)$, of G is the minimum cardinality of a dominating set of G. The maximum cardinality of a minimal dominating set of G is the *upper domination number*, $\Gamma(G)$, of G. A dominating set of G of cardinality $\gamma(G)$ is called a $\gamma(G)$-*set*. A vertex in G is *dominating* if it is adjacent to all other vertices of G. For more information on domination in graphs see the book [65].

Replacing closed neighborhoods with open neighborhoods leads one to the concept of total domination. In particular, if X and Y are subsets of vertices in a graph G, then the set X *totally dominates* the set Y in G if $Y \subseteq N(X)$. In particular, a vertex u *totally dominates* a vertex v if $v \in N(u)$. In the case that a set X totally dominates $V(G)$, then X is a *total dominating set* of G; that is, every vertex of G is adjacent to a vertex in X. Note that G has a total dominating set only if it has no isolated vertices. The *total domination number*, $\gamma_t(G)$, of a graph G with minimum degree at least 1 is the minimum cardinality of a total dominating set of G. A total dominating set of G of cardinality $\gamma_t(G)$ is called a $\gamma_t(G)$-*set*. For more information on total domination in graphs see the book [77].

A set of vertices A in a graph G is *independent* if any two distinct vertices in A are not adjacent, and the maximum size of an independent set in G is the *independence number*, $\alpha(G)$, of G. If A is an independent set in G, then the set B, where $B = V(G) - A$, is a *vertex cover* of G. Note that every edge in G is incident with at least one vertex in B. The minimum size of a vertex cover in G is the *vertex cover number*, $\beta(G)$, of G. A *matching* in G is a set of edges of G, no two of which share an endvertex. The maximum cardinality of a matching in G is the *matching number*, $\alpha'(G)$, of G.

A graph G is *connected* if there is a path joining every two vertices of G; otherwise, G is *disconnected*. A maximal connected subgraph of a graph is a *(connected) component*. The *distance* $d_G(u, v)$ between two vertices u and v in a connected graph G is the minimum length of a u, v-path in G. The *diameter* diam(G) of G is the maximum distance among all pairs of vertices of G. The *distance* between two sets S and T of vertices of G is the minimum distance between a vertex of S and a vertex of T; that is, $d_G(S, T) = \min\{d_G(u, v) : u \in S, v \in T\}$. A set $P \subseteq V(G)$ is a 2-*packing* in G if every two distinct vertices in P are at distance at least 3.

Let G and H be two graphs. In the four standard graph products $G * H$, where $* \in \{\circ, \boxtimes, \times, \square\}$, the vertex set of the product of graphs G and H is equal to $V(G) \times V(H)$. Two vertices (g_1, h_1) and (g_2, h_2) are adjacent in $V(G * H)$ if

- either $g_1 g_2 \in E(G)$ or ($g_1 = g_2$ and $h_1 h_2 \in E(H)$) holds, when $* = \circ$, and $G \circ H$ is the *lexicographic product* of G and H;

- $(g_1g_2 \in E(G)$ and $h_1 = h_2)$ or $(g_1 = g_2$ and $h_1h_2 \in E(H))$ or $(g_1g_2 \in E(G)$ and $h_1h_2 \in E(H))$ holds, when $* = \boxtimes$, and $G \boxtimes H$ is the *strong product* of G and H;
- $g_1g_2 \in E(G)$ and $h_1h_2 \in E(H)$ holds, when $* = \times$, and $G \times H$ is the *direct product* of G and H;
- $(g_1g_2 \in E(G)$ and $h_1 = h_2)$ or $(g_1 = g_2$ and $h_1h_2 \in E(H))$ holds, when $* = \square$, and $G \square H$ is the *Cartesian product* of G and H.

For more information on graph products see the book [63].

The *disjoint union* of two graphs G and H will be denoted with $G \sqcup H$. The disjoint union is an associative operation, and we will denote by nG the disjoint union of n copies of a graph G. The *join* $G + H$ of G and H is obtained from $G \sqcup H$ by adding all possible edges joining $V(G)$ and $V(H)$. The join is also an associative operation on graphs, and the join of two or more totally disconnected graphs is called a *complete multipartite graph*. The *corona* $G \odot H$ of graphs G and H is the graph obtained from G by adding a disjoint copy H_v of H for each vertex v of G and joining v to each vertex of H_v. In particular, $G \odot K_1$ is called the *corona* of G, also denoted cor(G) in the literature.

A subgraph Q of a graph G is a *k-clique* or simply a *clique* if Q is isomorphic to the complete graph K_k. A vertex $u \in V(G)$ is *simplicial* if $N_G(u)$ induces a clique. A pair of vertices u and v are *closed twins*, or simply called *twins*, in a graph G if $N_G[u] = N_G[v]$ and *open twins* if $N_G(u) = N_G(v)$. A vertex u is an *(open) twin vertex* if u and v are (open) twins for some vertex v. A graph is *twin-free* if it contains no twins, and *open twin-free* if it contains no open twins.

A graph G is *a chordal graph* if all induced cycles of G are triangles (that is, they induce K_3). A graph G is a *split graph* if $V(G)$ can be partitioned into a complete subgraph and an independent set. Note that trees and split graphs are chordal graphs. A *cograph* is a graph in which there is no induced path P_4. If H is a graph, and G contains no induced subgraph isomorphic to H, then G is an *H-free* graph. The graph $K_{n \times 2}$ obtained from the even-ordered complete graph K_{2n} by removing all edges of a perfect matching is the *cocktail-party graph*. Given a graph G, the *line graph* $L(G)$ of G is the graph with $V(L(G)) = E(G)$ and $ef \in E(L(G))$ if edges e and f are adjacent in G.

A *hypergraph* $\mathcal{H} = (V, F)$ consists of the non-empty set of vertices V and a family F of non-empty subsets of V called *edges*. The concept of hypergraph generalizes that of graph in which every edge has cardinality 2. A hypergraph in which all edges have the same cardinality k is a *k-uniform hypergraph*. (A graph is thus a 2-uniform hypergraph.) The *complete k-uniform hypergraph* on n vertices has all k-subsets of V as its edges. For a graph G, the *open (closed) neighborhood hypergraph*, abbreviated ONH (resp. CNH), of G is the hypergraph H with vertex set $V(H) = V(G)$ and with edge set $E(H) = \{N_G(x) : x \in V(G)\}$ (resp. $E(H) = \{N_G[x] : x \in V(G)\}$) consisting of the open (closed) neighborhoods of vertices in G.

If $n \in \mathbb{N}$, then we will use the notation $[n] = \{1, \ldots, n\}$, and let $[n]_0 = \{0, 1, \ldots, n\}$. For notation and graph theory terminology not defined herein, we in general follow [109], and for domination-related concepts we follow [65].

1.2   Domination and Total Domination Game

The domination game is played on a graph G by two players, *Dominator* and *Staller*, who take turns choosing a vertex from G. Whenever a vertex is chosen it must dominate at least one vertex not dominated by the vertices previously chosen. We call such a vertex a *playable* vertex. When either player chooses a vertex we call this a *move* in the game. For emphasis we will sometimes refer to a move as a *legal move*. When convenient we might also refer to the chosen vertex as a move. The game ends when no move is possible. Note that at this stage the set of selected vertices is a dominating set of G. The players have opposite goals. Dominator wishes that the game is ended in as few moves as possible, and Staller wishes to maximize the total number of moves.

If the first move is made by Dominator, then the game will be called a *D-game* and the sequence of moves will be denoted by $d_1, s_1, d_2, s_2, \ldots$ Otherwise (if Staller starts the game), the game will be called an *S-game*; in this case the sequence of moves will be denoted $s_1', d_1', s_2', d_2', \ldots$

The *game domination number*, $\gamma_g(G)$, of G is the number of moves in a D-game when both players follow a strategy to achieve their goals. The *Staller-start game domination number*, $\gamma_g'(G)$, of G is the number of moves in an S-game when both players follow a strategy to achieve their goals.

The *total domination game* is defined just as the domination game, except that whenever a player selects a new vertex in the course of the game, the selected vertex must **totally** dominate at least one vertex that was not **totally** dominated by vertices previously selected by the players. The *game total domination number* and the *Staller-start game total domination number* are denoted by $\gamma_{tg}(G)$ and $\gamma_{tg}'(G)$, respectively.

We first argue that the invariants γ_g, γ_g', γ_{tg}, and γ_{tg}' are well-defined. For this purpose consider the *game tree* of a domination game played on a graph G defined as follows. Its vertices are all the possible positions in the game, where a *position* is a sequence of legal moves played up to some point in a game played on G. The tree is rooted in the initial empty sequence λ. A position $(x_1, \ldots, x_{i+1})$, $i \geq 0$, is a child of a position $(x_1, \ldots, x_i)$. Note that $N[x_{i+1}] - N[\{x_1, \ldots, x_i\}] \neq \emptyset$. If $N[\{x_1, \ldots, x_k\}] = V(G)$, then the position $(x_1, \ldots, x_k)$ is *terminal* and represents a leaf of the game tree. In Figure 1.1 an example of a game tree is shown, where positions $(x_1, \ldots, x_i)$ are briefly written $x_1 \ldots x_i$. In the figure some branches of the tree are omitted due to symmetry. For instance, playing x' as the first move is equivalent/symmetric to playing x, hence the option x' is not drawn. Similarly, the first move y' is not shown. We do likewise in the rest of the game tree. Say, if z is

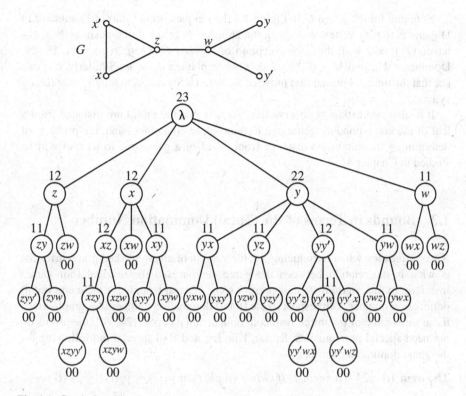

Fig. 1.1 Graph G and its game tree.

played as the first move, then playing y or y' as the second move is equivalent, and hence the (second) move y' is not drawn.

We next define the *value* of a position P in the game tree as an ordered pair of non-negative integers $(d(P), s(P))$ as follows. If P is a leaf, then set $(d(P), s(P)) = (0, 0)$. Otherwise, if $P_1, \ldots, P_\ell$ are the children of P, then

$$d(P) = 1 + \min_{i \in [\ell]} s(P_i) \quad \text{and} \quad s(P) = 1 + \max_{i \in [\ell]} d(P_i). \tag{1.1}$$

See Figure 1.1 again, where next to each position P the value $(d(P), s(P))$ is abbreviated as $d(P)s(P)$.

Note that the game tree contains all possible games played on a graph G and that any particular game corresponds to a path from the root λ to a leaf. Hence computing the values of the positions in the game tree in the reversed BFS order (starting from leaves), and having in mind (1.1), we infer that $(d(\lambda), s(\lambda)) = (\gamma_g(G), \gamma'_g(G))$. This in particular implies that γ_g and γ'_g are well-defined graph invariants. Similarly we can also build the game tree for the total domination game and conclude that also γ_{tg} and γ'_{tg} are well-defined.

Note that for the graph G in Figure 1.1 the unique optimal start of Dominator in
D-game is to play vertex w. Indeed, the children P_i of the initial position $P = (\lambda)$
have $s(P_i) = 2$ with the only exception of $P_4 = (w)$ where $s(w) = 1$. Hence,
Dominator will realize $\gamma_g(G) = 2$ only when playing $d_1 = w$. Similarly, one can
see that the unique optimal first move of Staller in a S-game is to play $s_1' = y$ (or, by
symmetry, $s_1' = y'$).

It is also interesting to observe that $\gamma(G)$ is just the minimum distance from a
leaf of the corresponding game tree to the root. On the other hand, the problem of
determining the maximum distance from a leaf of a game tree to its root will be
studied in Chapter 4.

1.3   Bounds in Terms of the (Total) Domination Number

A basic question when introducing a game version of an established graph invariant
is what is the relation between the game version and the original (non-game)
invariant. In this section, we thus present some general bounds for the game (total)
domination number with respect to the (total) domination number of a graph. Since
these were some of the initial results in both seminal papers [24, 70], the proofs do
not need special prerequisites. Brešar, Klavžar, and Rall showed the following for
the game domination number.

Theorem 1.1 [24, Theorem 1] *If G is a graph, then $\gamma(G) \leq \gamma_g(G) \leq 2\gamma(G) - 1$.*

Proof. When the D-game is played on a graph G, the set of vertices chosen by
Dominator and Staller together is a dominating set of G, implying that $\gamma(G) \leq
\gamma_g(G)$. To prove that $\gamma_g(G) \leq 2\gamma(G) - 1$, Dominator's strategy is to select an
arbitrary minimum dominating set S of G, order these vertices, and play the vertices
from the set sequentially according to this ordering, provided such a move is legal.
If a selected vertex in the set S is not playable, then Dominator simply considers the
next available vertex in the set. However, when Dominator exhausts the sequence
of vertices in S the graph is dominated, and hence no more moves are legal. Thus,
Dominator plays at most $|S|$ moves and Staller at most $|S| - 1$ moves. In this way,
Dominator can guarantee that the game finishes in at most $2|S| - 1 = 2\gamma(G) - 1$
moves.   □

Theorem 1.1 can be complemented by the following family of examples, which
show that all possible values can be attained. For an arbitrary positive integer k, and
any $r \in [k - 1]_0$, the graph $G_{k,r} = (r + 1)K_{1,k} \sqcup (k - r - 1)K_2$ has the property
$\gamma(G_{k,r}) = k$ and $\gamma_g(G_{k,r}) = k + r$.

If u and v are twins in a graph G, then u and v are dominated at the same time in
the game and at most one of u and v can be played in the game.

Observation 1.2 *If v is a twin vertex in a graph G, then $\gamma_g(G - v) = \gamma_g(G)$ and
$\gamma_g'(G - v) = \gamma_g'(G)$.*

In a manner similar to Theorem 1.1, the game total domination number can be bounded by the total domination number as shown by Henning, Klavžar, and Rall.

Theorem 1.3 [70, Theorem 3.1] *If G is an isolate-free graph, then*

$$\gamma_t(G) \le \gamma_{tg}(G) \le 2\gamma_t(G) - 1.$$

Moreover, given any integer $n \ge 2$ and any $0 \le \ell \le n - 1$ there exists a connected graph H such that $\gamma_t(H) = n$ and $\gamma_{tg}(H) = n + \ell$.

Let G be an isolate-free graph. By Theorem 1.1 and Theorem 1.3, we have $\gamma(G) \le \gamma_g(G) \le 2\gamma(G) - 1 \le 2\gamma_t(G) - 1 \le 2\gamma_{tg}(G) - 1$. Thus, Henning, Klavžar, and Rall [70] proved that the game domination number and the game total domination number are related as follows.

Theorem 1.4 [70, Theorem 3.2] *If G is an isolate-free graph, then $\gamma_g(G) \le 2\gamma_{tg}(G) - 1$.*

To see that Theorem 1.4 is close to being optimal, consider the following example given in [70]. For any $k \ge 2$, let G_k be the graph obtained from the complete graph on k vertices by attaching k leaves to each of its vertices. In Chapters 2 and 3 we will see two important general facts, known as the Continuation Principle and the Total Continuation Principle, which will allow us to assume what is an optimal first move for Dominator in either type of the game on G_k.

Consider first the total domination game played on G_k. We may assume (by the Total Continuation Principle) that the first move for Dominator (in an optimal game) is to play a vertex v of the k-clique. An optimal reply of Staller is to play a leaf neighbor of v because otherwise she could play no leaf in due course. After the first two moves both players must alternatively play all the $k - 1$ remaining vertices of the k-clique. Therefore, $\gamma_{tg}(G_k) = k + 1$.

Consider next the (usual) domination game played on G_k. We may assume (by the Continuation Principle) that the first move for Dominator (in an optimal game) is to play a vertex of the k-clique. Since each vertex of the k-clique has k leaf neighbors, Staller will be able to play a leaf as long as Dominator has not played all the vertices from the clique, implying that $\gamma_g(G_k) \ge 2k - 1$. On the other hand by Theorem 1.1, $\gamma_g(G_k) \le 2\gamma(G_k) - 1 = 2k - 1$. Thus,

$$\gamma_{tg}(G_k) = k + 1 \quad \text{and} \quad \gamma_g(G_k) = 2k - 1.$$

The game total domination number can be bounded by the domination number as follows.

Theorem 1.5 [70, Theorem 4.1] *If G is a graph such that $\gamma(G) \ge 2$, then $\gamma(G) \le \gamma_{tg}(G) \le 3\gamma(G) - 2$. Moreover, the bounds are sharp.*

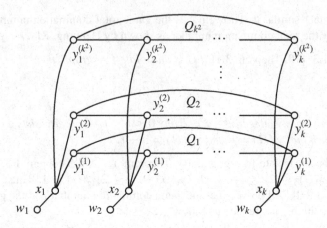

Fig. 1.2 The graph B_k.

To show sharpness of the upper bound in Theorem 1.5, the graph B_k, $k \geq 2$, is constructed in [70] as follows. For $i \in [k^2]$, let Q_i be a complete graph of order k with vertex set $\{y_1^{(i)}, \ldots, y_k^{(i)}\}$. Then take the disjoint union of these cliques, add vertices $x_1, \ldots, x_k$, and for $i \in [k]$, join x_i to the k^2 vertices $y_i^{(1)}, \ldots, y_i^{(k^2)}$. Finally, add a pendant edge to each vertex x_i and call the resulting leaf w_i. See Figure 1.2 for B_k. For further reference let $X = \{x_1, \ldots, x_k\}$, and for $i \in [k]$, let $Y_i = \{y_i^{(j)} : j \in [k^2]\}$.

We note that $\gamma(B_k) = k$ and in fact, X is a unique minimum dominating set of B_k. By Theorem 1.5, $\gamma_{tg}(B_k) \leq 3k - 2$. It therefore suffices for us to show that Staller has a strategy that will ensure the game to last $3k - 2$ moves. Her strategy is to play vertices from Y_1 as long as possible and to never play a vertex from X. Since as long as not all vertices from X are played, Staller can play inside Y_1, at least $k - 1$ moves are played on vertices from Y_1. Note next that during the game all vertices from X must be played in order to totally dominate the vertices w_i. In order to totally dominate the vertices from $X' = X - \{x_1\}$, an additional $k - 1$ moves are needed since X' is a 2-packing. Hence at least $(k-1) + k + (k-1) = 3k - 2$ moves are needed to complete the game, implying that $\gamma_{tg}(B_k) \geq 3k - 2$. Consequently, $\gamma_{tg}(B_k) = 3k - 2 = 3\gamma(B_k) - 2$.

Chapter 2
Domination Game

The domination game as defined in Chapter 1 was introduced by Brešar, Klavžar, and Rall in [24] as a game of perfect information played by two players who have opposite goals. As we have seen, two graphical invariants arise from the game depending on whether Dominator or Staller has the first move. Since these invariants are not based on finding a subset of vertices but rather on a vertex-by-vertex selection process, the proof techniques in game domination are quite different from those in ordinary domination theory. In the first part of this chapter, we introduce and illustrate the primary and the most important method of proof, namely the imagination strategy. Having a basic understanding of the imagination strategy is crucial to the theory, and readers who are new to the domination game are advised to master Sections 2.1 and 2.2 before proceeding into the meat of the book.

Section 2.3 is concerned with one of the first questions that arose upon the game's introduction, that of finding bounds on the game domination number in terms of the order of the underlying graph. In particular, we provide the proof by Henning and Kinnersley [68] and that of Bujtás [30], both of which show that $\gamma_g(G) \leq \frac{2n}{3}$ for a general graph G of order n with no isolated vertices. The conjectured upper bound of $\frac{3n}{5}$ is discussed in Section 2.4. In Section 2.5, we restrict attention to the domination game played on trees. It is rare that one can find the exact value of the game domination number, even in common classes of graphs. Section 2.6 covers several of these isolated instances where exact formulas are known. The conjecture that $\gamma_g(G) \leq \lceil \frac{n}{2} \rceil$ for a traceable graph G of order n is discussed next. This conjecture would be trivially true if the invariant γ_g were monotone on subgraphs. Results in Section 2.8 illustrate how far this is from being true in general. In analyzing the game played on a graph, it is often important, as the game progresses, to know what the effect is on the game domination number if certain vertices are already dominated. Thus the notions of criticality and stability of γ_g are introduced and studied in Section 2.9. Thereafter, we focus on the class of perfect graphs, those for which the domination number and the game domination number

© The Author(s), under exclusive license to Springer Nature Switzerland AG 2021
B. Brešar et al., *Domination Games Played on Graphs*, SpringerBriefs in
Mathematics, https://doi.org/10.1007/978-3-030-69087-8_2

are equal for all induced subgraphs. Since Staller's goal is to prolong the game, it perhaps seems intuitive that allowing her to start the game should give her an advantage in achieving this goal. In general this is not the case, and in Section 2.11, we present the connection between graphs in which Staller does not want to be the first to play and those graphs for which every vertex is an optimal first move for whichever player has the first move. We conclude this chapter with a presentation of the computational complexity of the domination game.

2.1   Continuation Principle

A *partially dominated graph* is a graph together with a declaration that some vertices are already dominated and need not be dominated in the rest of the game. Given a graph G and a subset S of vertices of G, we denote by $G|S$ the partially dominated graph with S as the set of declared vertices already dominated. We use $\gamma_g(G|S)$ (resp. $\gamma_g'(G|S)$) to denote the number of moves remaining in the game on $G|S$ under optimal play when Dominator (resp. Staller) has the next move. Kinnersley, West, and Zamani in [86, Lemma 2.1] presented the following key lemma, named the *Continuation Principle*.

Lemma 2.1 (Continuation Principle) *If G is a graph and $A, B \subseteq V(G)$ with $B \subseteq A$, then $\gamma_g(G|A) \leq \gamma_g(G|B)$ and $\gamma_g'(G|A) \leq \gamma_g'(G|B)$.*

Proof. Two games will be played in parallel, Game A on the partially dominated graph $G|A$ and Game B on the partially dominated graph $G|B$. The first of these will be the real game, while Game B will be imagined by Dominator. In Game A, Staller will play optimally while in Game B, Dominator will play optimally.

We claim that in each stage of the games, the strategy of Dominator as described below ensures that the set of vertices that are dominated in Game B is a subset of the set of vertices that are dominated in Game A. Since $B \subseteq A$, this is true at the start of the games. Suppose now that Staller has (optimally) selected a vertex u in Game A. Then by the induction assumption, the vertex u is a legal move in Game B because a new vertex v, which was dominated by u in Game A, is not yet dominated in Game B. Then Dominator copies the move of Staller by playing the vertex u in Game B. Dominator then replies with an optimal move in Game B. (In particular, if the D-game is being played and this is the first move in the game, then Dominator will choose an optimal move in Game B.) If this move is legal in Game A, Dominator plays it in Game A as well. Otherwise, if the game is not yet over, Dominator plays any other legal move in Game A. In either case, the set of vertices that are dominated in Game B is a subset of the set of vertices that are dominated in Game A which by induction also proves the claim.

We have thus proved that Game A finishes no later than Game B. Suppose that Game B lasted r moves. Because Dominator was playing optimally in Game B, it follows that $r \leq \gamma_g(G|B)$. On the other hand, because Staller was playing optimally in Game A and Dominator has a strategy to finish the game in r moves, we infer that $\gamma_g(G|A) \leq r$. Therefore,

$$\gamma_g(G|A) \le r \le \gamma_g(G|B),$$

and we are done if Dominator is the first to play. Note that in the above argument we did not assume who starts first, hence in both cases Game A will finish no later than Game B. Hence the conclusion holds for γ_g' as well; that is, $\gamma_g'(G|A) \le \gamma_g'(G|B)$.

□

As a consequence of the Continuation Principle, whenever x and y are legal moves for Dominator in the domination game and $N[x] \subseteq N[y]$, we may assume that Dominator will play y instead of x. Furthermore, if it was Staller's turn, we may assume that Staller will play x instead of y. Further as an important consequence of the Continuation Principle, we have the fundamental property of the domination game that the number of moves in the D-game and the S-game when played optimally can differ by at most 1; see [24, Theorem 6] and [86, Corollary 4.1].

Theorem 2.2 *For any graph G, we have $|\gamma_g(G) - \gamma_g'(G)| \le 1$.*

Proof. Consider the D-game and let v be the first move of Dominator. Let $A = N[v]$ and let $B = \emptyset$, and consider the partially dominated graph $G|A$. We note that $G|B = G$. Since v is an optimal first move of Dominator, we note further that $\gamma_g(G) = 1 + \gamma_g'(G|A)$. By the Continuation Principle, $\gamma_g'(G|A) \le \gamma_g'(G|B) = \gamma_g'(G)$. Therefore,

$$\gamma_g(G) \le \gamma_g'(G|A) + 1 \le \gamma_g'(G) + 1.$$

By a parallel argument, consider the S-game and let v be the first move of Staller. As before, let $A = N[v]$ and $B = \emptyset$, and consider the partially dominated graph $G|A$. Since v is an optimal first move of Staller, we note that $\gamma_g'(G) = 1 + \gamma_g(G|A)$. By the Continuation Principle, $\gamma_g(G|A) \le \gamma_g(G|B) = \gamma_g(G)$, implying that

$$\gamma_g'(G) \le \gamma_g(G|A) + 1 \le \gamma_g(G) + 1.$$

This completes the proof of Theorem 2.2.   □

We remark that the proof of Lemma 2.1 could be modified to work on some, but not all, variants of possible domination games. There are several variants of the domination games, such as the independent domination game and the connected domination game, in which the Continuation Principle does not hold. These two games, among others, will be discussed in Chapter 5. In such games, the number of moves in the D-game and the S-game when played optimally can often differ by an arbitrarily large constant.

Using the Continuation Principle, Kinnersley, West, and Zamani in [86] showed that the length of the D-game in a partially dominated forest with no isolated vertex can never exceed the length of the S-game.

Theorem 2.3 [86, Theorem 4.6] *If F is a partially dominated forest with no isolated vertex, then $\gamma_g(F) \le \gamma_g'(F)$.*

We say that a graph G *realizes* the pair (k, ℓ) if $\gamma_g(G) = k$ and $\gamma_g'(G) = \ell$. By Theorem 2.2, for a given k, we have $\ell \in \{k - 1, k, k + 1\}$. All these possibilities are realizable with the exception of the pair $(2, 1)$. That this pair is not realizable follows by the fact that $\gamma_g'(G) = 1$ if and only if G is a complete graph. We next demonstrate the realizability of the other pairs.

- Pairs (k, k).
  Complete graphs K_n realize $(1, 1)$. For $k \geq 2$, the graph $C_4 \sqcup (k - 2)K_2$ realizes (k, k). Indeed, no matter which player is the first to select a vertex of C_4, exactly two vertices will be played on C_4 and, clearly, one move will be played on each K_2.
- Pairs $(k, k + 1)$.
  For $k \geq 1$ the graph $P_3 \sqcup (k - 1)K_2$ realizes the pair $(k, k + 1)$. In the D-game Dominator's optimal first move is the center of P_3, while in the S-game Staller first selects a leaf of P_3.
- Pairs $(k, k - 1)$.
  Let $k \geq 3$ be odd. Consider the graph $G_k = C_6 \sqcup (k - 3)K_2$. If $k = 3$, then $G_3 = C_6$. In the D-game, Staller can reply to any (equivalent) move d_1 with a neighbor of d_1, thus forcing $\gamma_g(C_6) = 3$. On the other hand, in the S-game Dominator can reply with the vertex at distance 3 from s_1', yielding $\gamma_g'(C_6) = 2$. Hence G_3 realizes $(3, 2)$. Now assume that $k \geq 5$. Then the number of disjoint copies of K_2 in G_k is even. From the above argument on C_6, neither of the players wishes to be the first to play a vertex of G_k from the C_6 component. This implies that in the D-game, Dominator will be the first to play on this component, while in the S-game Staller will be forced to play on C_6 first. In conclusion, G_k realizes $(k, k - 1)$ for $k \geq 3$ and k odd.

  Let $k \geq 4$ be even. It was observed in [86, Proposition 4.3] that $P_4 \square K_2$ realizes the pair $(4, 3)$. (The reader is invited to perform a corresponding, relatively simple case analysis.) Let now $H_k = (P_4 \square K_2) \sqcup (k/2 - 2)C_4$. Again using the fact that no matter which player is the first to select a vertex of C_4, exactly two vertices will be played on it; H_k realizes $(k, k - 1)$ for $k \geq 4$ and k even.

Examples of 2-connected, twin-free graphs that realize all these pairs will be given in Section 2.6.

We close the section with the following application of Theorem 2.2.

Corollary 2.4 *If a partially dominated graph $G|S$ realizes $(k, k - 1)$, then for any playable vertex u in $G|S$, the graph $G|(S \cup N[u])$ realizes $(k - 2, k - 1)$.*

Proof. Since u is a legal, but perhaps not optimal, move in $G|S$, we have from Dominator's and Staller's point of view, respectively,

$$k = \gamma_g(G|S) \leq 1 + \gamma_g'(G|(S \cup N[u])) \text{ and } k - 1 = \gamma_g'(G|S) \geq 1 + \gamma_g(G|(S \cup N[u])).$$

From these inequalities and Theorem 2.2, we get:

$$k - 1 \leq \gamma'_g(G|(S \cup N[u])) \leq \gamma_g(G|(S \cup N[u])) + 1 \leq k - 1.$$

Hence the equality holds everywhere, which shows that $\gamma_g(G|(S \cup N[u])) = k - 2$ and $\gamma'_g(G|(S \cup N[u])) = k - 1$. □

2.2 Imagination Strategy

One of the most useful tools for proving bounds on the domination game is the imagination strategy. The strategy was first introduced and applied in the seminal paper [24], and we already used an instance of it in the proof of the Continuation Principle. With its frequent use in the proofs of game domination theory, it seems that the imagination strategy was born just for this theory. Nevertheless, without giving it this name, the idea of imagination strategy appeared in the context of coloring games [6, 113], and may well be applicable also in other games on graphs.

In general, when a domination game is played on a graph, one of the players imagines another game is played at the same time, usually on a copy of the same or slightly modified graph. We then speak of a real game and an imagined game. For the imagined game, the optimal number of moves k is known, and hence if Dominator (resp. Staller) is the player who imagines the parallel game, then he (resp. she) has a strategy that ensures the imagined game takes at most (at least) k moves. The basic strategy in the real game is to copy each move of the opponent to the imagined game, respond in the imagined game by an optimal move, and finally copy back this move to the real game. While the two games are played, two possible problems must be resolved one way or another: some of the moves that are legal in the imagined game need not be legal in the real game, and some of the moves of the opponent in the real game need not be legal in the imagined game. The overall aim is to ensure that the number of moves in the real game is bounded by the number of moves in the imagined game. For purposes of illustration, suppose that the real game is being played on G and Dominator is playing the imagined game on H. Suppose now that when both games are finished, s vertices were selected in the real game played on G and t vertices were selected in the imagined game played on H. Since Dominator may not play optimally on G (but Staller does), we have $\gamma_g(G) \leq s$. Similarly, since Dominator plays optimally on H, we have $\gamma_g(H) \geq t$. If, for instance, $s \leq t$ holds, then we can conclude that $\gamma_g(G) \leq s \leq t \leq \gamma_g(H)$.

For another application of the imagination strategy, we next prove that on certain subgraphs the game domination number is hereditary. A subgraph H of a graph G is a *guarded* subgraph if for every vertex x of G, there exists a vertex $y \in V(H)$ such that $N_G[x] \cap V(H) \subseteq N_G[y] \cap V(H)$. Such a vertex y is called a *guard* of x in H. As shown by Brešar, Klavžar, Košmrlj, and Rall [23], the domination number of a guarded subgraph H is at most that of the original graph G; that is, $\gamma(H) \leq \gamma(G)$. We now have the following result.

Theorem 2.5 [23, Theorem 3.2] *If H is a guarded subgraph of a graph G, then*

$$\gamma_g(H) \leq \gamma_g(G) \quad \text{and} \quad \gamma_g'(H) \leq \gamma_g'(G).$$

Proof. Let the D-game be played on the graph G. We now present a strategy for Staller that will force at least $\gamma_g(H)$ moves to be made in this game. At the same time that the game is being played on G, Staller imagines a D-game on H. Dominator will play optimally on G while Staller will play optimally on H. Each move of Dominator in the real game will be used by Staller to determine a move for Dominator in the imagined game on H. Staller will respond optimally in H and copy this move into the real game. The details of how these moves are made by Staller will be described in the remainder of the proof. At any point in the game (after a move has been made or imagined by either player in both games), we let S_{real} denote the set of vertices belonging to H that are dominated in the real game on G, and we let S_{im} denote the set of vertices belonging to H that are dominated in the imagined game. The key property, ensured by Staller's strategy, is the following:

($\mathcal{P}$) After each move, made by either player, $S_{\text{real}} \subseteq S_{\text{im}}$.

At the beginning of the game, property ($\mathcal{P}$) is certainly true since then $S_{\text{real}} = S_{\text{im}} = \emptyset$. We proceed to show that ($\mathcal{P}$) holds throughout the game. Suppose that v is a vertex, played by Dominator at some point in the (real) game. Let S_{im}' be the set of vertices of H that were dominated in the imagined game before Dominator played the vertex v. Similarly, let S_{real}' denote the set of vertices of H that were dominated in the real game before Dominator played v. If $N[v] \cap (V(H) - S_{\text{im}}') \neq \emptyset$, then Staller imagines that Dominator played a guard u of v with respect to G. Note that this is a legal move for Dominator in the imagined game since $N[v] \cap V(H) \subseteq N[u] \cap V(H)$. In addition, S_{real} is a subset of S_{im} since $S_{\text{real}}' \subseteq S_{\text{im}}'$, and hence ($\mathcal{P}$) holds after this move of Dominator. On the other hand, if $N[v] \cap (V(H) - S_{\text{im}}') = \emptyset$, then an arbitrary vertex w from H is chosen by Staller for a move of Dominator in the imagined game, as long as it is a legal move in this game. Again it is obvious that property ($\mathcal{P}$) is true after this move since $S_{\text{real}} = S_{\text{real}}' \subseteq S_{\text{im}}' \subseteq S_{\text{im}}$. If $N[v] \cap (V(H) - S_{\text{im}}') = \emptyset$ and no such legal move w is available in the imagined game, then Staller has ensured that the real game lasts at least as long as the imagined game since the imagined game was finished before Dominator played v.

Now, when it is Staller's turn, she responds in the imagined game by playing a vertex according to her optimal strategy of making the imagined game last as long as possible. She copies this move to the real game. That is, if her choice is a playable vertex x, then she plays x also in the real game. Note that by property ($\mathcal{P}$), this move will be legal also in the real game, and it is clear that ($\mathcal{P}$) is preserved after her move since S_{real} and S_{im} are enlarged by the set $N[x] \cap V(H)$. Now, since Staller is playing optimally in the imagined game, it will last at least $\gamma_g(H)$ steps. By ($\mathcal{P}$), the real game will not end before the imagined game is over; hence, the real game will also last at least $\gamma_g(H)$ steps. Since Staller was not necessarily playing optimally in the real game and Dominator was, we infer that the real game lasted at most $\gamma_g(G)$ moves. Therefore, $\gamma_g(H) \leq \gamma_g(G)$.

Finally, when the S-game is played in a graph G, the same strategy as in the D-game can be used by Staller, from which we derive $\gamma_g'(H) \leq \gamma_g'(G)$. □

As a direct consequence of Theorem 2.5, we have the following result.

Corollary 2.6 [23, Corollary 3.4] *If $H_1, \ldots, H_m$ are guarded subgraphs in a graph G, and $d_G(V(H_i), V(H_j)) \geq 3$ whenever $i \neq j$, then*

$$\gamma_g(G) \geq \sum_{i=1}^{m} \gamma_g(H_i) - m + 1.$$

In the next result by Klavžar, Košmrlj, and Schmidt [88], we bound from below the game domination number in terms of the diameter. Its proof is based on Theorem 2.5 on guarded subgraphs.

Proposition 2.7 [88, Proposition 2.5] *If G is a graph, then*

(a) $\mathrm{diam}(G) \leq \begin{cases} 2\gamma_g(G); & \gamma_g(G) \text{ odd}, \\ 2\gamma_g(G) - 1; & \gamma_g(G) \text{ even}; \end{cases}$

(b) $\mathrm{diam}(G) \leq 2\gamma_g'(G) - 1$.

Moreover, the bounds are tight.

Proof. Let P be a shortest path in G of length $\mathrm{diam}(G)$. As P is a geodesic, a vertex $v \in V(G) - V(P)$ can be adjacent to at most three vertices of P, and the distance between the first and the last such vertex on P (if they exist) is at most 2. It follows that P is guarded in G. Hence by Theorem 2.5, $\gamma_g(G) \geq \gamma_g(P)$ and $\gamma_g'(G) \geq \gamma_g'(P)$. From the latter inequality and the Staller-start game domination number of paths (see Theorem 2.33), we get

$$\gamma_g'(G) \geq \gamma_g'\left(P_{\mathrm{diam}(G)+1}\right) = \left\lceil \frac{\mathrm{diam}(G) + 1}{2} \right\rceil \geq \frac{\mathrm{diam}(G) + 1}{2},$$

which proves (b).

Let $\gamma_g(G)$ be odd. Suppose by way of contradiction that $\mathrm{diam}(G) \geq 2\gamma_g(G) + 1$ and let P' be a shortest path of order $2\gamma_g(G) + 2$. Then $\gamma_g(G) \geq \gamma_g(P') = \lceil (2\gamma_g(G) + 2)/2 \rceil = \gamma_g(G) + 1$, a contradiction. Hence $\mathrm{diam}(G) \leq 2\gamma_g(G)$ holds when $\gamma_g(G)$ is odd. The argument for even $\gamma_g(G)$ proceeds similarly.

The tightness of the bounds can be demonstrated by considering appropriate paths in Theorem 2.33: if $k \geq 1$, then $\mathrm{diam}(P_{4k}) = 4k - 1 = 2\gamma_g(P_{4k}) - 1$; $\mathrm{diam}(P_{4k+3}) = 4k + 2 = 2\gamma_g(P_{4k+3})$; and $\mathrm{diam}(P_{2k}) = 2k - 1 = 2\gamma_g'(P_{2k}) - 1$. □

2.3 Upper Bounds and the Bujtás Discharging Method

In 2013, Kinnersley, West, and Zamani in [86] were the first to prove a general upper bound on the game domination number of an isolate-free graph in terms of its order. They proved that if G is a general isolate-free graph G of order n, then $\gamma_g(G) \leq \lceil \frac{7}{10}n \rceil$. This upper bound was first improved independently by Bujtás [30] and Henning and Kinnersley [68] to $\frac{2}{3}n$. Due to the relative simplicity of these proofs which illustrate the type of arguments used in the domination game, we present both proofs as a gentle introduction to the domination game. (A stronger upper bound will also be presented, but without proof.)

For this purpose, we consider a partially dominated graph G with a set of vertices predominated. In such a graph, certain vertices and edges are irrelevant in that their presence or absence does not affect the game. We call a vertex of G *saturated* if it and all of its neighbors have already been dominated. We define the *reduced graph* G' corresponding to G to be the graph obtained from G by deleting all edges joining dominated vertices. The removal of such edges makes no difference to the outcome of the game, and $\gamma_g(G') = \gamma_g(G)$ and $\gamma_g'(G') = \gamma_g'(G)$. We note that in a reduced graph, all neighbors of a dominated vertex must be undominated. We are now in a position to present the proof of the following general upper bound on the game domination number given in [68].

Theorem 2.8 [68, Theorem 2.2] *Let G be a partially dominated graph that does not contain an isolated undominated vertex. If G has order n with d dominated vertices and s saturated vertices, then*

$$\gamma_g(G) \leq \frac{1}{3}(2n - s - d) \quad and \quad \gamma_g'(G) \leq \frac{1}{3}(2n - s - d + 1).$$

Proof. Without loss of generality, we may assume that G is a reduced graph. Thus, G has $s \geq 0$ saturated vertices and these vertices are isolated in G, and the remaining vertices of G, if any, are not isolated. We proceed by induction on $n - s$. If $n - s = 0$, then every vertex of G is saturated, implying that $n = s = d$ and $\gamma_g(G) = \gamma_g'(G) = 0 = (2n - s - d)/3$. This establishes the base case. Let $n - s > 0$, and so G has at least one non-trivial component (that is, a component with at least two vertices). For the inductive hypothesis, assume that if G' is a reduced graph of order n with d' dominated vertices and s' saturated vertices where $s' > s$, then $\gamma_g(G') \leq (2n - s' - d')/3$ and $\gamma_g'(G') \leq (2n - s' - d' + 1)/3$.

Suppose that G has $k \geq 1$ non-trivial components, each isomorphic to K_2. In this case, $n - s = 2k$ and, since each K_2-component contains at least one undominated vertex, $d \leq s + k$. Moreover, in the remainder of the game, exactly one move gets made in each component. Thus,

$$\gamma_g(G) = \gamma_g'(G) = k = \frac{1}{3}(2(s + 2k) - s - (s + k)) \leq \frac{1}{3}(2n - s - d),$$

and so both bounds hold. Hence we may assume that G contains at least one component of order at least 3, for otherwise the desired result holds.

Suppose Staller plays first. Staller's move must dominate at least one new vertex and saturates at least one new vertex (in particular, the vertex Staller played). Let G' denote the resulting reduced graph of order n with d' dominated vertices and s' saturated vertices. We note that G' contains no isolated undominated vertex, since every non-trivial component of G' has at least one undominated vertex, and all undominated vertices in G' have the same degree as in G. Thus,

$$
\begin{aligned}
\gamma_g'(G) &= 1 + \gamma_g(G') \\
&\le 1 + \tfrac{1}{3}(2n - s' - d') \\
&\le 1 + \tfrac{1}{3}(2n - (s+1) - (d+1)) \\
&= \tfrac{1}{3}(2n - s - d + 1),
\end{aligned}
$$

where the first equality follows from optimality of Staller's move, the second inequality follows from the induction hypothesis applied to G', and the third inequality follows from our earlier observation that $d' \ge d + 1$ and $s' \ge s + 1$.

Suppose next that Dominator plays first. Let C be an arbitrary component of G of order at least 3. We have two cases to consider. In each case, we specify Dominator's move, and we let G' denote the resulting reduced graph of order n with d' dominated vertices and s' saturated vertices.

Case 1. C has a leaf Let v be any leaf of C and let u be its neighbor. If both v and u are undominated, then Dominator plays u; this dominates and saturates both u and v. Thus, $s' \ge s + 2$ and $d' \ge d + 2$. Now,

$$
\begin{aligned}
\gamma_g(G) &\le 1 + \gamma_g'(G') \\
&\le 1 + \tfrac{1}{3}(2n - s' - d' + 1) \\
&\le 1 + \tfrac{1}{3}(2n - (s+2) - (d+2) + 1) \\
&= \tfrac{1}{3}(2n - s - d),
\end{aligned}
$$

where the first inequality holds because Dominator does at least as well playing optimally as he does by playing on u, and the second inequality follows from the induction hypothesis applied to G'.

Suppose instead the vertex v is undominated but the vertex u is dominated. Since the component C has order at least 3 and there is no edge between dominated vertices in C, this implies that u has some undominated neighbor w different from v. Dominator plays u, thereby dominating v and w, and saturating u and v. Thus once again two new vertices are dominated and saturated, and so $s' \ge s+2$ and $d' \ge d+2$ and analogously as before, $\gamma_g(G) \le \tfrac{1}{3}(2n - s - d)$.

The only remaining possibility is that v is dominated but u is not. If u has an undominated neighbor w, then Dominator plays w, which dominates u and w, and saturates v and w. Once again, $s' \ge s + 2$ and $d' \ge d + 2$, and $\gamma_g(G) \le \tfrac{1}{3}(2n - s - d)$ holds. Finally, suppose that u has no undominated neighbors. The choice of

C implies that u has at least one neighbor other than v, say t. Dominator plays t, which dominates u and saturates t and v. Since u had no undominated neighbors, this move also saturates u, implying that $s' \geq s + 3$ and $d' \geq d + 1$ and

$$
\begin{aligned}
\gamma_g(G) &\leq 1 + \gamma_g'(G') \\
&\leq 1 + \tfrac{1}{3}(2n - s' - d' + 1) \\
&\leq 1 + \tfrac{1}{3}(2n - (s + 3) - (d + 1) + 1) \\
&= \tfrac{1}{3}(2n - s - d).
\end{aligned}
$$

Case 2. C has no leaves Suppose firstly that some undominated vertex v in C has at least two undominated neighbors. In this case, Dominator plays v, which dominates at least three vertices, namely v and its undominated neighbors, and saturates v, implying that $s' \geq s + 1$ and $d' \geq d + 3$ and

$$
\begin{aligned}
\gamma_g(G) &\leq 1 + \gamma_g'(G') \\
&\leq 1 + \tfrac{1}{3}(2n - s' - d' + 1) \\
&\leq 1 + \tfrac{1}{3}(2n - (s + 1) - (d + 3) + 1) \\
&= \tfrac{1}{3}(2n - s - d).
\end{aligned}
$$

Hence we may assume that no undominated vertex in C has two undominated neighbors, for otherwise the desired bound holds. Suppose that some undominated vertex v in C has exactly one undominated neighbor. Let u be the undominated neighbor of v. By assumption, u has no other undominated neighbors. Dominator plays u; this dominates and saturates both u and v, implying that $s' \geq s + 2$ and $d' \geq d + 2$ and, as before, $\gamma_g(G) \leq \tfrac{1}{3}(2n - s - d)$. Hence we may assume that C contains no adjacent undominated vertices.

Since G' contains no isolated undominated vertex, C contains at least one pair of adjacent vertices, at least one of which must be dominated; let v be a dominated vertex in C. Since C has no leaves, v must have at least two neighbors. Recall that in the reduced graph no edge joins two dominated vertices. In particular, every neighbor of v is undominated. Let u and w be two undominated neighbors of v; by assumption, neither u nor w has any undominated neighbors. Dominator plays v, which dominates u and w, and saturates u, v, and w, implying that $s' \geq s + 3$ and $d' \geq d + 2$ and

$$
\begin{aligned}
\gamma_g(G) &\leq 1 + \gamma_g'(G') \\
&\leq 1 + \tfrac{1}{3}(2n - s' - d' + 1) \\
&\leq 1 + \tfrac{1}{3}(2n - (s + 3) - (d + 2) + 1) \\
&< \tfrac{1}{3}(2n - s - d).
\end{aligned}
$$

This completes the proof of Theorem 2.8.   □

Theorem 2.8 yields the following result in the special case when $d = s = 0$.

Corollary 2.9 [68, Corollary 2.3] *If G is an isolate-free graph of order n, then*

$$\gamma_g(G) \leq \frac{2}{3}n \quad and \quad \gamma_g'(G) \leq \frac{2n+1}{3}.$$

We provide next an alternative proof of the $\frac{2}{3}n$-bound on the game domination number using an ingenious approach first adopted by Bujtás [29, 30]. Her approach, which we call the *Bujtás discharging method*, is to color the vertices with three (or more) colors that reflect three (or more) different types of vertices and to associate a weight with each vertex. At any stage of the game, if D denotes the set of vertices played to date where initially $D = \emptyset$, we define a *colored-graph* with respect to the played vertices in the set D as a graph in which every vertex is colored with one of three colors, namely white, blue, or red, according to the following rules:

- A vertex is colored *white* if it is not dominated by D.
- A vertex is colored *blue* if it is dominated by D but has a neighbor not dominated by D.
- A vertex is colored *red* if it and all its neighbors are dominated by D.

In a colored-graph, the only playable vertices are those that are white or have a white neighbor since a played vertex must dominate at least one new vertex. In particular, no red vertex is playable. Once a vertex is colored red, it plays no role in the remainder of the game, and edges joining two blue vertices play no role in the game. Therefore, we may assume a colored-graph contains no red vertices and has no edge joining two blue vertices. The resulting colored-graph corresponds to the *residual graph* which is obtained from the partially dominated graph by removing every edge between two dominated vertices and deleting the saturated vertices. We note that the degree of a white vertex in the residual graph remains unchanged from its degree in the original graph. We also note that since G contains no isolated vertices at the beginning of the game, this property remains valid for each residual graph throughout the game. We next associate a weight with each colored vertex as given in Table 2.1.

We denote the *weight* of a vertex v in the residual graph G by $w(v)$. For a subset $S \subseteq V(G)$ of vertices of G, the *weight* of S is the sum of the weights of the vertices in S, denoted $w(S)$. The *weight* of G, denoted $w(G)$, is the sum of the weights of the vertices in G; that is, $w(G) = w(V(G))$. We define the *value* of a playable vertex as the decrease in weight resulting from playing that vertex.

Theorem 2.10 *If G is an isolate-free colored-graph of order n, then* $\gamma_g(G) \leq \frac{1}{3}w(G).$

Table 2.1 The weights of vertices according to their color.

Color of vertex	Weight of vertex
White	2
Blue	1
Red	0

Proof. Let G be an isolate-free colored-graph of order n. We say that Dominator can *achieve his* 3-*target* if he can play a sequence of moves guaranteeing that on average the weight decrease resulting from each played vertex in the game is at least 3. In order to achieve his 3-target, Dominator must guarantee that a sequence of moves $m_1, \ldots, m_k$ are played, starting with his first move m_1, and with moves alternating between Dominator and Staller such that if w_i denotes the decrease in weight after move m_i is played, then

$$\sum_{i=1}^{k} w_i \geq 3k, \qquad (2.1)$$

and the game is completed after move m_k. In the discussion that follows, we analyze how Dominator can achieve his 3-target. Before any move of Dominator, the game is in one of the following two phases:

- *Phase* 1, if there exists a legal move of value at least 4.
- *Phase* 2, if every legal move has value at most 3.

We proceed further with the following claims.

Claim 2.11 *Every legal move in a residual graph decreases the total weight by at least* 2.

Proof. Every legal move in a colored-graph is a white vertex or a blue vertex with at least one white neighbor. Let v be a legal move in a residual graph. Suppose that v is a white vertex. When v is played, the vertex v is recolored red while each white neighbor of v, if any, is recolored blue or red, and each blue neighbor of v, if any, remains blue or is recolored red, implying that the weight decrease resulting from playing v is at least 2. Suppose that v is a blue vertex. In this case, every neighbor of v is white and v has at least one white neighbor. When v is played, the vertex v is recolored red while each (white) neighbor of v is recolored blue or red. Hence, the total weight decrease resulting from playing v is at least 2.   (□)

Claim 2.12 *The game is in Phase* 2 *if and only if every component of the residual graph is a* K_2-*component with one vertex colored white and the other colored blue.*

Proof. If every component of the residual graph is a K_2-component with one vertex colored white and the other colored blue, then every vertex is a legal move and when played recolors exactly one blue vertex red and exactly one white vertex red, implying that every legal move has weight 3. This establishes the sufficiency. To prove the necessity, suppose that the game is in Phase 2.

If there exists a white vertex v with at least two white neighbors, then when v is played, the vertex v is recolored red while each white neighbor of v is recolored blue or red, implying that the weight decrease resulting from playing v is at least 4, contradicting the supposition that the game is in Phase 2. Hence, every white vertex has at most one white neighbor.

If there are two adjacent white vertices, then all other neighbors of these two white vertices are colored blue. Hence when one of these white vertices is played, both white vertices are recolored red, decreasing the weight by at least 4, a contradiction. Hence, all neighbors of a white vertex are blue.

If there exists a blue vertex v with at least two white neighbors, then when v is played, the vertex v is recolored red and each white neighbor of v is recolored red, implying that the weight decrease resulting from playing v is at least 5, a contradiction. Hence, each blue vertex has exactly one white neighbor, and each white vertex has exactly one blue neighbor and no white neighbor. Thus, every component of the residual graph is a K_2-component with one vertex colored white and the other colored blue. $\hspace{1cm}$ ($\square$)

We now return to the proof of Theorem 2.10. We show that Dominator can achieve his 3-target by following a greedy strategy. Thus, at each stage of the game, Dominator plays a (greedy) move that decreases the weight by as much as possible. By Claim 2.11, every move of Staller decreases the weight by at least 2. Hence, whenever Dominator plays a vertex that decreases the weight by at least 4, his move, together with Staller's response, decreases the weight by at least 6. Therefore, we may assume that at some stage the game enters Phase 2, for otherwise Inequality (2.1) is satisfied upon completion of the game and Dominator can achieve his 3-target. By Claim 2.12, this implies that every component of the residual graph is a K_2-component with one vertex colored white and the other colored blue. Thus, every legal move in the remaining part of the game has value exactly 3, and so once again Inequality (2.1) is satisfied upon completion of the game and Dominator can achieve his 3-target. $\hspace{1cm}$ $\square$

As a consequence of Theorem 2.10, we have the following slightly stronger result than Corollary 2.9.

Corollary 2.13 [30, Proposition 2] *If G is an isolate-free graph of order n, then*

$$\gamma_g(G) \leq \frac{2}{3}n \quad and \quad \gamma_g'(G) \leq \frac{2}{3}n.$$

Proof. Coloring the vertices of G with the color white, we produce a colored-graph in which every vertex is colored white. In particular, we note that G has n white vertices and has weight $w(G) = 2n$. By Theorem 2.10, $\gamma_g(G) \leq \frac{1}{3}w(G) = \frac{2}{3}n$. In the S-game, the first move of Staller is a white vertex with at least one white neighbor, which when played decreases the weight by at least 3. Thus if G' is the colored-graph after Staller's first (optimal) move, then G' is isolate-free and $w(G') \leq w(G) - 3 = 2n - 3$. Thus,

$$\gamma_g'(G) = 1 + \gamma_g(G') \leq 1 + \frac{1}{3}w(G') \leq \frac{2}{3}n,$$

where the first equality follows from optimality of Staller's move, and the last inequality follows from Theorem 2.10 applied to the colored-graph G'. $\hspace{1cm}$ $\square$

We remark that the result of Corollary 2.13 in the special case when G is chordal was proved earlier in [86, Proposition 5.3].

2.4 $\frac{3}{5}$-Conjectures

A significant part of the interest in the domination game arose from attempts to improve the upper bound on the game domination number in terms of the order of a given graph. In 2013, Kinnersley, West, and Zamani [86] posted the so-called $\frac{3}{5}$-Conjecture, which has attracted considerable attention and has yet to be fully settled. In this section, we discuss the $\frac{3}{5}$-Conjecture and its current status. We remark that there are two $\frac{3}{5}$-Conjectures: one for isolate-free forests, and one for general isolate-free graphs. We state both conjectures.

Conjecture 2.14 [86] If G is an isolate-free forest of order n, then $\gamma_g(G) \leq \frac{3}{5}n$.

Conjecture 2.15 [86] If G is an isolate-free graph of order n, then $\gamma_g(G) \leq \frac{3}{5}n$.

To distinguish these two conjectures, we refer to Conjecture 2.14 for isolate-free forests as the $\frac{3}{5}$-**Forest Conjecture**, and we refer to Conjecture 2.15 for general isolate-free graphs as the $\frac{3}{5}$-**Graph Conjecture**. It is not known whether the $\frac{3}{5}$-Forest Conjecture implies the $\frac{3}{5}$-Graph Conjecture.

2.4.1 The $\frac{3}{5}$-Forest Conjecture

In this section, we discuss the history and current status of the $\frac{3}{5}$-Forest Conjecture. In 2013, Kinnersley, West, and Zamani [86, Theorem 5.5] showed that the $\frac{3}{5}$-Forest Conjecture holds when G is an isolate-free forest of caterpillars. In 2013, Brešar, Klavžar, Košmrlj, and Rall [22] verified the $\frac{3}{5}$-Forest Conjecture for all trees on at most 20 vertices, and listed those meeting the conjectured bound with equality (when $n = 20$, there are only ten such trees).

In 2015, using the Bujtás discharging method, Bujtás [29, Theorem 2] proved that if G is an isolate-free forest, then $\gamma_g(G) \leq \frac{5}{8}n(G)$. By using the same technique, Bujtás [29, Theorem 1] proved the $\frac{3}{5}$-Forest Conjecture for isolate-free forests in which no two leaves are at distance 4 apart. This result was extended in 2016 by Schmidt [106] to the larger class of weakly $S(K_{1,3})$-free forests, where $S(K_{1,3})$ is the graph obtained from a star $K_{1,3}$ by subdividing every edge once and where a weakly $S(K_{1,3})$-free forests is an isolate-free forest without induced $S(K_{1,3})$ whose leaves are leaves of the forest as well. The big breakthrough came in 2016 when Marcus and Peleg announced they had proven the $\frac{3}{5}$-Forest Conjecture in an unpublished manuscript [95].

It remains an open problem to characterize the isolate-free forests that achieve equality in the $\frac{3}{5}$-Forest Conjecture. In 2013, Brešar, Klavžar, Košmrlj, and Rall [22] presented a construction that yields an infinite family of trees that attain the bound in the $\frac{3}{5}$-Forest Conjecture. A much larger, but simpler, construction of extremal trees was subsequently presented in 2017 by Henning and Löwenstein in [73]. In order to explain this construction, they defined the notion of a 2-wing as follows.

Definition 2.16 A tree T is a *2-wing* if T has maximum degree at most 4 with no vertex of degree 3, and with the vertices of degree 2 in T precisely the support vertices of T, except for one vertex of degree 2 in T. This exceptional vertex of degree 2 in T, which is not a support vertex, is the *gluing vertex* of T.

As remarked in [73], the smallest 2-wing is a path on five vertices, with its central vertex as the gluing vertex. A 2-wing with gluing vertex v is illustrated in Figure 2.1.

Definition 2.17 A tree T belongs to the family $\mathcal{T}$ if T is obtained from $k \geq 1$ vertex-disjoint 2-wings by adding $k - 1$ edges between the gluing vertices.

It is shown in [73] that the family of trees constructed in [22] is a proper subfamily of trees in the family $\mathcal{T}$. An example of a tree that belongs to the family $\mathcal{T}$ but does not belong to the family of trees constructed in [22] is shown in Figure 2.2.

Theorem 2.18 [73, Theorem 6] *If $T \in \mathcal{T}$, then $\gamma_g(T) = \gamma_g'(T) = \frac{3}{5}n(G)$.*

The following conjecture was posed in 2017 by Henning and Löwenstein in [73] and has yet to be resolved.

Conjecture 2.19 [73, Conjecture 1] If F is an isolate-free forest of order n satisfying $\gamma_g(F) = \frac{3}{5}n$, then every component of F belongs to the family $\mathcal{T}$.

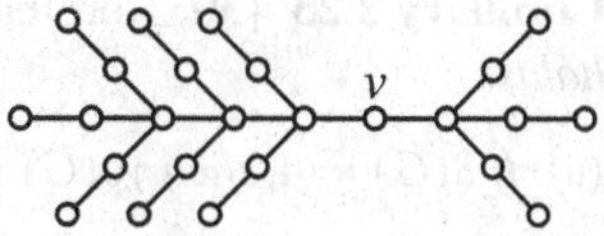

Fig. 2.1 A 2-wing with gluing vertex v.

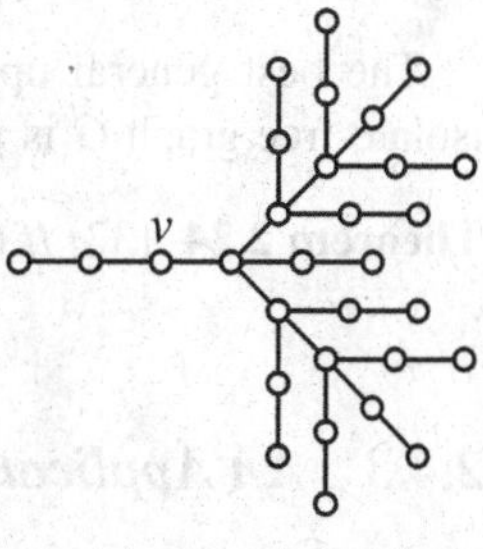

Fig. 2.2 A tree $T \in \mathcal{T}$.

2.4.2 The $\frac{3}{5}$-Graph Conjecture

We next discuss progress on the $\frac{3}{5}$-Graph Conjecture. In 2016, Henning and Kinnersley [68] proved the $\frac{3}{5}$-Graph Conjecture when the graph G has minimum degree at least 2.

Theorem 2.20 [68, Theorem 2.7] *The $\frac{3}{5}$-Graph Conjecture is true for all graphs with minimum degree at least 2.*

However, the $\frac{3}{5}$-Graph Conjecture has yet to be settled in general for graphs that contain vertices of degree 1. When the minimum degree is at least 3, Bujtás [30] obtained an upper bound smaller than $0.5574n < \frac{3}{5}n$ on the game domination number of graphs of order n. More precisely, she proved the following remarkable results using the Bujtás discharging method.

Theorem 2.21 [30, Theorem 3] *If G is a graph of order n with $\delta(G) \geq 3$, then*

$$\gamma_g(G) \leq \frac{34}{61}n \quad and \quad \gamma_g'(G) \leq \frac{34n - 27}{61}.$$

More generally, Bujtás [30] proved the following result.

Theorem 2.22 [30, Theorem 4] *If G is a graph of order n with $\delta(G) \geq \delta \geq 4$, then*

$$\gamma_g(G) \leq \left(\frac{15\delta^4 - 28\delta^3 - 129\delta^2 + 354\delta - 216}{45\delta^4 - 195\delta^3 + 174\delta^2 + 174\delta - 216} \right) n.$$

As an immediate consequence of Theorem 2.22, we have the following result.

Corollary 2.23 [30, Corollary 5] *If G is a graph of order n, then the following holds:*

(a) *If $\delta(G) = 4$, then $\gamma_g(G) \leq \frac{37}{72}n < 0.5139n$.*

(b) *If $\delta(G) \geq 5$, then $\gamma_g(G) \leq \frac{2102}{4377}n < 0.4803n$.*

The best general upper bound to date on the game domination number of an isolate-free graph G is $\gamma_g(G) \leq \frac{5}{8}n(G)$ given by Bujtás.

Theorem 2.24 [32] *If G is an isolate-free graph, then $\gamma_g(G) \leq \frac{5}{8}n(G)$.*

2.4.3 An Application of the Bujtás Discharging Method

As demonstrated above, the Bujtás discharging method has been the central tool so far for attacking the $\frac{3}{5}$-Conjectures. In addition, the method turned out to be applicable also for the problem described in this subsection.

A central theme in domination theory is to find upper bounds for the domination number of graphs with given minimum degree in terms of their order. Back in 1962, Ore [99] observed that if $\delta(G) \geq 1$, then $\gamma(G) \leq n(G)/2$. Blank [7] and (much later but independently) McCuaig and Shepherd [96] proved that if $\delta(G) = 2$ and G is not one of seven sporadic graphs (C_4, and six graphs on seven vertices), then $\gamma(G) \leq 2n(G)/5$. In 1996, Reed [102] proved that if $\delta(G) = 3$, then $\gamma(G) \leq 3n(G)/8$. The result is sharp since cubic graphs of order 8 with domination number 3 exist, but as soon as a cubic graph has at least ten vertices, then $\gamma(G) \leq 4n(G)/11$ holds, a result due to Kostochka and Stodolsky [91]. The same bound also holds for graphs of minimum degree 4 as proved by Sohn and Xudong [107]. Applying the Bujtás discharging method a simpler proof of this theorem is given in [28]. In the same paper and using the same method, the following result is proved.

Theorem 2.25 [28, Theorem 1] *If G is a graph with $\delta(G) = 5$, then $\gamma(G) \leq n(G)/3$.*

The best earlier known bound $\gamma(G) \leq \frac{2671}{7766} < 0.344$ for graphs G with $\delta(G) = 5$ was proved by Bujtás and Klavžar in [42], also by applying Bujtás discharging method. Moreover, in the same paper, best known bounds for graphs of minimum degree from the set $\{6, 7, \ldots, 50\}$ were derived. In particular, the following holds.

Theorem 2.26 [42, Corollary 1]

(i) If G is a graph with $\delta(G) = 6$, then

$$\gamma(G) \leq \frac{1702}{5389} n(G) < 0.3159 \, n(G).$$

(ii) If G is a graph with $\delta(G) = 7$, then

$$\gamma(G) \leq \frac{389701}{1331502} n(G) < 0.2927 \, n(G).$$

This method has been applied also for the 2-domination number of a graph. (The 2-*domination number* of a graph G is the smallest cardinality of a set $S \subseteq V(G)$ such that every vertex in $V(G) - S$ has at least two neighbors in S.) Bujtás and Jaskó used the discharging method in [41] for proving upper bounds on the 2-domination number of G in terms of $n(G)$ when $\delta(G)$ is fixed. They improve the best earlier bounds for any $6 \leq \delta(G) \leq 21$.

2.5 Domination Game on Trees

Brešar, Klavžar, and Rall [25] studied the domination game in trees. They obtained the following lower bound on the game domination number of a tree.

Theorem 2.27 [25, Theorem 2.2] *If T is a tree of order n and maximum degree Δ, then*

$$\gamma_g(T) \geq \left\lceil \frac{2n}{\Delta + 3} \right\rceil - 1,$$

and this bound is asymptotically best possible.

A family of caterpillars presented in [25] shows that the bound is asymptotically optimal. Notably, for $s - 1 \geq t$, let $T(s, t)$ be the tree obtained from a path P_t by attaching $s - 1$ pendant edges to each vertex of the path. The resulting tree is a caterpillar of order $n = st$ and maximum degree $\Delta = s + 1$. As observed in [25], $\gamma_g(T(s,t)) = 2t - 1$. We note that

$$\left\lceil \frac{2n}{\Delta + 3} \right\rceil - 1 = \left\lceil \frac{2st}{s + 4} \right\rceil - 1,$$

which for a fixed t and for s sufficiently large, is arbitrarily close to $2t - 1$.

As noted in Theorem 1.1, for a given graph G, the smallest possible value for $\gamma_g(G)$ (and also for $\gamma_g'(G)$) is the domination number of G. We say that G is a γ_g-*minimal graph* if the equality $\gamma_g(G) = \gamma(G)$ holds. Similarly, if $\gamma_g'(G) = \gamma(G)$, then G is called a γ_g'-*minimal graph*. Simple examples of γ_g-minimal graphs are P_4, C_4, complete graphs, and complete multipartite graphs in which either some color classes consist of a single vertex or where at least one of the color classes has cardinality 2. Any complete multipartite graph with no color class of cardinality 1 is a γ_g'-minimal graph; other small examples include C_5, C_6, and C_7. For disconnected graphs, the situation can be quite subtle. The graphs K_1 and C_6 are each γ_g'-minimal, but $K_1 \sqcup C_6$ realizes $(3, 4)$ and thus is not γ_g'-minimal. On the other hand, C_6 is not γ_g-minimal, but $K_1 \sqcup C_6$ is γ_g-minimal. This shows that care must be taken in analyzing disjoint unions.

The situation is more tractable for forests. Using Theorem 2.3, we see that $\gamma(F) \leq \gamma_g(F) \leq \gamma_g'(F)$ holds for any forest F, and hence a γ_g'-minimal forest is also γ_g-minimal. It then follows that a forest is γ_g'-minimal if and only if each of its components is γ_g'-minimal. Nadjafi-Arani, Siggers, and Soltani [97] gave a characterization of the class of γ_g-minimal forests and the γ_g'-minimal forests.

Theorem 2.28 [97, Theorem 4.3] *A forest F is γ_g'-minimal if and only if each component of F is a path of order 1, 2, or 4.*

To characterize the γ_g-minimal forests requires more work. First, for each $i \in [5]$ we specify a tree T_i and for each a distinguished vertex, x_i, which we will call an *attachment* vertex. These trees and their attachment vertices are as follows:

- $T_1 = K_1$ and x_1 is the only vertex.
- $T_2 = K_2$ and x_2 is either vertex.
- $T_3 = P_5$ and x_3 is a leaf.
- $T_4 = P_4$ and x_4 is a support vertex.
- $T_5 = P_3 \odot K_1$ and x_5 is a support vertex of degree 2.

Theorem 2.29 [97, Theorem 4.1] *A tree T is γ_g-minimal if and only if $T = K_1$ or T is constructed from a vertex x and a finite collection of vertex disjoint copies of trees chosen (repetition allowed) from $\{T_1, T_2, T_3, T_4, T_5\}$, where at least one copy of T_1 is chosen, by making x adjacent to each of the attachment vertices in the collection.*

Using the characterization from Theorem 2.29, the authors of [97] were able to completely describe the γ_g-minimal forests.

Corollary 2.30 [97, Corollary 4.4] *A forest is γ_g-minimal if and only if it is γ_g'-minimal, or it is the disjoint union of a γ_g'-minimal forest and a γ_g-minimal tree.*

On the other hand, by Theorem 1.1 we know that for a given graph G the largest possible value of $\gamma_g(G)$ is $2\gamma(G) - 1$. The graphs G for which $\gamma_g(G) = 2\gamma(G) - 1$ holds were investigated by Xu, Li in [111] and Klavžar in [112] and named γ_g-*maximal*. In [112, Theorem 5.2], starlike γ_g-maximal trees are characterized, where a tree is *starlike* if it contains exactly one vertex of degree at least 3. The characterization displays an explicit list of starlike γ_g-maximal trees and requires quite some effort to deduce it. An additional large family of γ_g-maximal trees is given in [112, Theorem 3.3]. Moreover, some starlike γ_g-maximal graphs which are not trees were also obtained. To describe them, let $\text{Supp}(G)$ denote the set of support vertices of G, that is, vertices adjacent to leaves. Then we have the following result.

Theorem 2.31 [112, Theorem 3.1] *Let G be a connected graph of order at least 3. If $\text{Supp}(G)$ forms a dominating set and there are at least $\lceil \log_2 \gamma(G) \rceil + 1$ pendant vertices adjacent to each vertex of $\text{Supp}(G)$, then G is γ_g-maximal.*

2.6   Exact Values of the Game Domination Number

Exact values for non-trivial classes of graphs are very rare. Proofs to determine the exact value of the game domination number, even for relatively simple graph classes, tend to be very complex.

2.6.1   Paths and Cycles

The following result for paths and cycles was first reported by Kinnersley, West, and Zamani in an unpublished manuscript. An alternative proof with additional information on the game in paths and cycles was given by Košmrlj [93]. Along with the formulas, he also gives optimal strategies for both players.

Theorem 2.32 [93, Theorem 2.2] *If $n \geq 3$, then*

$$\gamma_g(C_n) = \begin{cases} \left\lceil \frac{n}{2} \right\rceil - 1; \ n \equiv 3 \ (\text{mod } 4) \,, \\ \\ \left\lceil \frac{n}{2} \right\rceil ; \qquad otherwise, \end{cases}$$

and

$$\gamma_g'(C_n) = \begin{cases} \left\lceil \frac{n-1}{2} \right\rceil - 1; \ n \equiv 2 \ (\text{mod } 4) \,, \\ \\ \left\lceil \frac{n-1}{2} \right\rceil ; \qquad otherwise. \end{cases}$$

Theorem 2.33 [93, Theorem 2.4] *If $n \geq 1$, then*

$$\gamma_g(P_n) = \begin{cases} \left\lceil \frac{n}{2} \right\rceil - 1; \ n \equiv 3 \ (\text{mod } 4) \,, \\ \\ \left\lceil \frac{n}{2} \right\rceil ; \qquad otherwise, \end{cases}$$

and $\gamma_g'(P_n) = \left\lceil \frac{n}{2} \right\rceil$.

2.6.2 Powers of Cycles

If G is a graph and $n \geq 1$, then the nth power G^n of G is the graph with $V(G^n) = V(G)$, where two vertices are adjacent in G^n if and only if their distance in G is at most n. The following result of Bujtás, Klavžar, and Košmrlj [43] extends the result on cycles.

Theorem 2.34 [43, Theorem 9] *If $n \geq 1$ and $N \geq 3$, then*

$$\gamma_g(C_N^n) = \begin{cases} \left\lceil \frac{N}{n+1} \right\rceil ; \qquad N \bmod (2n+2) \in \{0, 1, \ldots, n+1\} \,, \\ \\ \left\lceil \frac{N}{n+1} \right\rceil - 1; \ N \bmod (2n+2) \in \{n+2, \ldots, 2n+1\} \,. \end{cases}$$

Moreover, if $n \geq 1$ and $N \geq 2n+1$, then

$$\gamma_g'(C_N^n) = \begin{cases} \left\lceil \frac{N}{n+1} \right\rceil ; \qquad N \bmod (2n+2) \in \{0\} \,, \\ \\ \left\lceil \frac{N}{n+1} \right\rceil - 1; \ N \bmod (2n+2) \in \{1, \ldots, n+1, 2n+1\} \,, \\ \\ \left\lceil \frac{N}{n+1} \right\rceil - 2; \ N \bmod (2n+2) \in \{n+2, \ldots, 2n\} \,. \end{cases}$$

2.6.3 Coronas

The next result shows that to determine the exact value of the game domination number in the simple family of the corona of paths is also very subtle.

Theorem 2.35 [92, Theorem 4.1] *If $k \geq 1$, then*

$$\gamma_g(P_k \odot K_1) = k + \left\lceil \frac{k-7}{10} \right\rceil \quad and \quad \gamma'_g(P_k \odot K_1) = k + \left\lceil \frac{k-2}{10} \right\rceil.$$

2.6.4 Disjoint Union of Special Graphs

Domination game played on the disjoint union of paths and cycles was considered by Ruksasakchai, Onphaeng, and Worawannotai [104]. They provided exact values for the game domination number and for the Staller-start game domination number of an arbitrary graph, each component of which is either a path or a cycle. The formulas are somewhat involved, and, in particular, the values depend on the lengths of paths and cycles modulo 4.

Worawannotai and Chantarachada [110] investigated the game domination number in the following class of graphs. First, a graph G is a *chain of complete graphs* $K_{n_1}, \ldots, K_{n_k}$ if G can be obtained from $K_{n_1}, \ldots, K_{n_k}$ by identifying a vertex q_i in K_{n_i} with a vertex t_{i+1} in $K_{n_{i+1}}$, for all $i \in [k-1]$, such that $t_i \neq q_i$ for all i. Second, a graph G is a *cycle of complete graphs* $K_{n_1}, \ldots, K_{n_k}$ if G can be obtained from the chain of complete graphs $K_{n_1}, \ldots, K_{n_k}$ by identifying a vertex in $V(K_{n_k}) - V(K_{n_{k-1}})$ with a vertex in $V(K_{n_1}) - V(K_{n_2})$. Finally, a graph G is in the class CC if each component of G is either a chain of complete graphs or a cycle of complete graphs.

Let G be a graph in the class CC. Let $M_{m_1}, \ldots, M_{m_r}$ be the components of G, which are the chains of m_i complete graphs (if such components exist), and let $N_{n_1}, \ldots, N_{n_s}$ be the components of G, which are the cycles of n_i complete graphs (if such components exist). Note that at least one of the integers among r and s is positive. By $a(G)$ we denote the number of those components of G that are chains of m_i complete graphs, where $m_i \equiv 2 \pmod 4$ or $m_i \equiv 3 \pmod 4$. Similarly, by $b(G)$ we denote the number of components of G that are cycles of n_i complete graphs such that $n_i \equiv 2 \pmod 4$.

Theorem 2.36 [110, Theorem 3.1] *If $G = M_{m_1} \sqcup \cdots \sqcup M_{m_r} \sqcup N_{n_1} \sqcup \cdots \sqcup N_{n_s}$ is a graph in the class CC, then*

$$\gamma_g(G) = \sum_{i=1}^{r} \left(m_i - \left\lfloor \frac{m_i}{4} \right\rfloor \right) + \sum_{i=1}^{s} \left(n_i - \left\lfloor \frac{n_i + 2}{4} \right\rfloor \right) + \left\lfloor \frac{b(G) - a(G)}{2} \right\rfloor,$$

and

$$\gamma_g'(G) = \sum_{i=1}^{r} \left(m_i - \left\lfloor \frac{m_i}{4} \right\rfloor \right) + \sum_{i=1}^{s} \left(n_i - \left\lfloor \frac{n_i+2}{4} \right\rfloor \right) + \left\lceil \frac{b(G)-a(G)}{2} \right\rceil .$$

2.6.5 Graphs with Game Domination Numbers 2 or 3

Klavžar, Košmrlj, and Schmidt [88] characterized several classes of graphs with small game domination number. Note first that $\gamma_g(G) = 1$ if and only if $\Delta(G) = n(G) - 1$ and that $\gamma_g'(G) = 1$ if and only if G is a complete graph. To characterize graphs with game domination number equal to 2, let us say that a *homogeneous clique* Q in a graph G is a clique in which every two vertices are twins, that is, $N_G[u] = N_G[v]$ holds for every $u, v \in V(Q)$.

Proposition 2.37 [88, Proposition 3.6] *If G is a connected graph, then the following hold:*

(i) $\gamma_g(G) = 2$ *if and only if* $\Delta(G) < n(G) - 1$ *and there exists a vertex* $v \in V(G)$ *such that* $V(G) - N[v]$ *induces a homogeneous clique in* G.
(ii) $\gamma_{g'}(G) = 2$ *if and only if G is not complete and every vertex lies in a dominating set of order* 2.

If $\gamma_g(G) = 2$ or $\gamma_g'(G) = 2$, then diam$(G) \leq 3$ by Proposition 2.7. A characterization of graphs with $\gamma_g(G) = 2$ and diam$(G) = 3$ can be easily deduced from Proposition 2.37(i). To describe the structure of graphs G with $\gamma_{g'}(G) = 2$ and diam$(G) = 3$, we define the following concept. A connected graph G is a *gamburger*, if G is obtained from the disjoint union of non-empty subgraphs Q_1, T_1, T_2, and Q_2, where Q_1 and Q_2 are cliques with no edges between them. Further, for $i \in [2]$ there is a join between Q_i and T_i and there are no edges between Q_i and T_{3-i}. Finally, for $i \in [2]$ and for every vertex $x \in V(T_i)$, there exists a vertex $x' \in V(T_{3-i}) \cup V(Q_{3-i})$ such that $V(T_1) \cup V(T_2) \subseteq N[x] \cup N[x']$. See Figure 2.3 where a gamburger is schematically presented.

Theorem 2.38 [88, Theorem 3.11] *A graph G has $\gamma_g'(G) = 2$ and diam$(G) = 3$ if and only if G is a gamburger.*

In [88] the class of graphs G with $\gamma_g(G) = 3$ and diam$(G) = 6$, and the class of graphs G with $\gamma_{g'}(G) = 3$ and diam$(G) = 5$ are also characterized, both in two different ways. The first class can be recognized in time $O(n(G)m(G) + \Delta(G)^3)$, see [87, Theorem 3.2].

Fig. 2.3 Gamburger (with cliques Q_1 and Q_2).

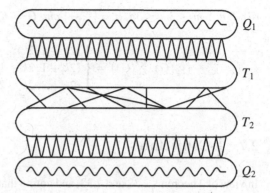

2.6.6 2-Connected Realizations

For $k \geq 2$, let $G_{k,k}$ be the graph obtained from a cycle C_{2k} by adding a new vertex and joining it to two vertices at distance 2 apart on the cycle. For $k \geq 2$, let $G_{k,k+1}$ be the graph obtained from a cycle C_{2k} by adding two new vertices u and v and joining both of them to two specified adjacent vertices on the cycle. We note that u and v are open twins (of degree 2) in $G_{k,k+1}$. For $k \geq 2$, let $G_{2k,2k-1}$ be the graph obtained from the disjoint union of a cycle C_{4k+2} and a prism $P_3 \,\square\, K_2$ by selecting two adjacent vertices u and v on the cycle, and two adjacent vertices u' and v' of degree 2 in the prism, and adding the edges uu' and vv'. For $k \geq 1$, let $G_{2k+1,2k}$ be the cycle C_{4k+2}.

With these definitions, we are ready to give examples of 2-connected, twin-free graphs that realize all possible pairs, except $(2, 1)$, as announced in Section 2.1. The assertions (a), (b), and (c) of the following theorem were given in [92, Theorems 2.2, 2.3, 3.3], while the last assertion (d) follows from Theorem 2.32.

Theorem 2.39 [92] *The following assertions hold:*

(a) *If $k \geq 2$, then $\gamma_g(G_{k,k}) = k$ and $\gamma_g'(G_{k,k}) = k$.*
(b) *If $k \geq 2$, then $\gamma_g(G_{k,k+1}) = k$ and $\gamma_g'(G_{k,k+1}) = k + 1$.*
(c) *If $k \geq 2$, then $\gamma_g(G_{2k,2k-1}) = 2k$ and $\gamma_g'(G_{2k,2k-1}) = 2k - 1$.*
(d) *If $k \geq 1$, then $\gamma_g(G_{2k+1,2k}) = 2k + 1$ and $\gamma_g'(G_{2k+1,2k}) = 2k$.*

2.6.7 Some Hamming Graphs

Proposition 2.40 [15, Proposition 5.1] *Let m and n be non-negative integers.*

(i) *If $n \geq 2m$, then*

$$\left(\gamma_g(K_{m+1} \,\square\, K_{n+1}), \gamma_g'(K_{m+1} \,\square\, K_{n+1})\right) = (2m + 1, 2m).$$

(ii) If $n + m \le 3m \le 3n$, $k = \lfloor \frac{n+m}{3} \rfloor$ *and* $G = K_{m+1} \,\square\, K_{n+1}$, *then*

$$\left(\gamma_g(G), \gamma_g'(G) \right) = \begin{cases} (2k+1, 2k); & n + m \equiv 0 \,(\mathrm{mod}\ 3), \\ (2k+1, 2k+1); & n + m \equiv 1 \,(\mathrm{mod}\ 3), \\ (2k+2, 2k+2); & n + m \equiv 2 \,(\mathrm{mod}\ 3). \end{cases}$$

2.7 $\frac{1}{2}$-Conjectures

In view of Theorems 2.32 and 2.33, and noting that paths and cycles are the simplest graphs with a hamiltonian path and a hamiltonian cycle, several years ago Rall proposed the following conjecture (explicitly stated for the first time in [82]) that strengthens the $\frac{3}{5}$-Graph Conjecture for the class of traceable graphs (that is, for the graphs containing hamiltonian paths).

Conjecture 2.41 If G is a traceable graph, then $\gamma_g(G) \le \left\lceil \frac{1}{2} n(G) \right\rceil$.

If H is a spanning subgraph of G, then the domination number of H is at least the domination number of G. However, Brešar, Klavžar, and Rall showed in [25] that the game domination number does not have this same type of behavior on subgraphs. In fact, as we describe in Section 2.8, for any positive integer k there exists a connected graph G and a spanning tree T of G such that $\gamma_g(G) \ge \gamma_g(T) + k$. In particular, Conjecture 2.41 does not follow from Theorems 2.32 and 2.33. Note that C_5 and C_6 both show that the above conjecture, if true, is sharp.

Several papers, whose results lend support to the truth of Conjecture 2.41, have appeared. James, Klavžar, and Vijayakumar [82] proved the following result. Recall that a split graph is one for which $V(G)$ can be partitioned into a complete graph, say K, and an independent set I. We assume that notation in this discussion.

Theorem 2.42 [82, Theorems 3.1 and 3.3] *If G is a connected split graph of order at least 2, then* $\gamma_g(G) \le \left\lfloor \frac{n(G)}{2} \right\rfloor$ *and* $\gamma_g'(G) \le \left\lfloor \frac{n(G)+1}{2} \right\rfloor$.

Therefore, any split graph that has a hamiltonian path satisfies Conjecture 2.41. James et al. also characterized the split graphs G of even order such that $\gamma_g(G) = \left\lfloor \frac{n(G)}{2} \right\rfloor$. Suppose that G is such a split graph. Their results imply that every vertex in K is adjacent to at least one leaf in I. Consequently G is not hamiltonian, and if G has a hamiltonian path, then $G = P_4$. They also observed that to prove Conjecture 2.41 for an arbitrary graph G, it is sufficient to restrict attention to those G with a hamiltonian path and for which $(n(G) + 2)/4 < \gamma(G) \le \lceil n(G)/3 \rceil$. This restriction on the domination number of G follows from Theorem 1.1 and the fact that a graph with a hamiltonian path has domination number at most $\lceil n(G)/3 \rceil$.

Recall that an edge cut of a connected graph G is a set, F, of edges of G such that $G - F$ is disconnected. The edge connectivity of G is the cardinality of a smallest edge cut and is denoted $\kappa'(G)$. In [89], Klavžar, and Rall proved the following result.

Theorem 2.43 [89, Corollary 3.2] *If G is a connected graph and C is a minimum edge cut of G, then $\gamma_g(G) \leq \gamma_g(G - C) + 2\kappa'(G)$.*

As an application of Theorem 2.43, the authors constructed the following infinite family of traceable graphs that satisfy Conjecture 2.41. Let $k \geq 15$ and let H be any traceable graph of order at most $3k - 42$. Let u and v be any two vertices of a complete graph K_k and let a and b be the ends of a hamiltonian path in H. The graph $K_k(H)$ is now obtained from the disjoint union of H and K_k by adding the edges au and bv. The set $\{au, bv\}$ is a minimum edge cut of $K_k(H)$, and as was shown in [89, Theorem 5.1], $\gamma_g(K_k(H)) \leq \left\lceil \frac{1}{2}n(K_k(H)) \right\rceil$. In addition to the above, Theorem 2.34 also provides evidence in support of Conjecture 2.41 in those cases where G is hamiltonian.

Further progress on Conjecture 2.41 was reported by Bujtás, Iršič, Klavžar, and Xu in [40]. In one of the main results of the paper, the conjecture was verified for unicyclic traceable graphs. This class of graphs contains (i) cycles, (ii) graphs obtained by attaching a path to a vertex of a cycle, and (iii) graphs obtained by attaching two disjoint paths to adjacent vertices of a cycle. The conjecture was further verified for several additional classes of traceable graphs, including particular Halin trees and graphs obtained from a cycle by connecting two nonadjacent vertices of the cycle [40, Proposition 5.1].

A possible approach to (dis)prove Conjecture 2.41 is the following. Let P be a hamiltonian path of G. Then the conjecture holds when the game is played on P. Hence, adding edges to P one by one, until G is reached, while keeping the game domination number below $\lceil n(G)/2 \rceil$, would yield the conjecture. Examples were found, however, where this desired monotonicity property does not hold. More precisely, we have $\gamma_g(P_{11}) = 5$, and the same value is achieved for all possibilities of adding one or two edges to the path P_{11}. But when three edges are added, it can happen that the game domination number increases to $6 = \lceil \frac{11}{2} \rceil$. In this respect, several computer experiments were performed but no counterexample found.

In [39], Conjecture 2.41 was proved for diameter 2 graphs. More precisely, in [39, Theorem 4.1] it is proved that if $\text{diam}(G) = 2$, then $\gamma_g(G) \leq \left\lceil \frac{n(G)}{2} \right\rceil - \left\lfloor \frac{n(G)}{11} \right\rfloor$. Moreover, the bound is sharp: if $\text{diam}(G) = 2$ and $n(G) \leq 10$, then $\gamma_g(G) = \left\lceil \frac{n(G)}{2} \right\rceil$ if and only if G is one of seven sporadic graphs with $n(G) \leq 6$ or the Petersen graph [39, Theorem 3.1].

Conjecture 2.41 is further valid on the class of line graphs.

Theorem 2.44 [36, Theorem 4.12] *If G is a traceable line graph, then $\gamma_g(G) \leq \left\lceil \frac{n(G)}{2} \right\rceil$.*

Based on the above developments, knowing that if $\delta(G) \geq 2$, then $\gamma_g(G) \leq \frac{3}{5}n(G)$ holds, and being aware of no minimum degree 2 graphs G for which $\gamma_g(G) > \left\lceil \frac{1}{2}n(G) \right\rceil$ would hold, the following conjecture was posed.

Conjecture 2.45 [36, Conjecture 1.2] If $\delta(G) \geq 2$, then $\gamma_g(G) \leq \left\lceil \frac{1}{2}n(G) \right\rceil$.

Provided a counterexample to Conjecture 2.45 will be found, the following weaker version of it has been proposed.

Conjecture 2.46 [36, Conjecture 6.2] There exists a constant $c < 3/5$ such that every graph G with $\delta(G) \geq 2$ and order n at least 6 satisfies $\gamma_g(G) \leq cn$.

In [36], it is proved that both Conjectures 2.41 and 2.45 hold true for claw-free cubic graphs. It is further proved that $\gamma_g(G) \leq \left\lceil \frac{11}{20} n(G) \right\rceil$ holds for every claw-free graph G of minimum degree at least 2.

2.8 Domination Games on Subgraphs

By an intricate application of the Continuation Principle and the imagination strategy, Brešar, Dorbec, Klavžar, and Košmrlj [14] proved the following result.

Theorem 2.47 [14, Theorem 2.1] *If e is an edge of a graph G, then*

$$|\gamma_g(G) - \gamma_g(G - e)| \leq 2 \quad and \quad |\gamma_g'(G) - \gamma_g'(G - e)| \leq 2.$$

The situation is different for the operation of vertex removal, where the (Staller-start) game domination number can increase by an arbitrary amount after this operation is performed. If H is an arbitrary graph with $\gamma_g(H) = r$ and $\gamma_g'(H) = s$, and G is the graph obtained from H by adding a vertex v adjacent to all vertices of H, then $\gamma_g(G) = 1$ and $\gamma_g(G - v) = r$, while $\gamma_g'(G) \leq 2$ and $\gamma_g'(G - v) = s$. Thus in contrast to Theorem 2.47, we have the following result.

Observation 2.48 [14] *If v is a vertex of a graph G, then the differences $\gamma_g(G - v) - \gamma_g(G)$ and $\gamma_g'(G - v) - \gamma_g'(G)$ can be arbitrarily large.*

On the other hand, the reversed relation between $\gamma_g(G)$ and $\gamma_g(G - v)$ is the same as for edge deletion, yet the proof is fairly straightforward.

Theorem 2.49 [14, Theorem 3.1] *If v is a vertex of a graph G, then*

$$\gamma_g(G) - \gamma_g(G - v) \leq 2 \quad and \quad \gamma_g'(G) - \gamma_g'(G - v) \leq 2.$$

Proof. Clearly, $\gamma_g'(G|N_G[v]) = \gamma_g'((G - v)|N_G(v))$ holds in an arbitrary graph G and for any vertex $v \in V(G)$ (and the symmetric equality holds for the game domination number too). Suppose that Dominator plays v as his first move in the D-game. This might not be an optimal first move for Dominator, but by the Continuation Principle we get

$$\gamma_g(G) \le 1 + \gamma_g'(G|N_G[v]) = 1 + \gamma_g'((G-v)|N_G(v)) \le 1 + \gamma_g'(G-v).$$

Theorem 2.2 further implies

$$1 + \gamma_g'(G-v)) \le \gamma_g(G-v) + 2,$$

which establishes the first inequality. For the second inequality we essentially use the same sequence of arguments.   □

Several non-trivial constructions of graphs are given in [14] to demonstrate that all possible values in Theorem 2.47 and Theorem 2.49 are realizable.

James, Dorbec, and Vijayakumar [81] proved that if G is from the class of no-minus graphs (see the definition in Section 2.11), then not all values in Theorem 2.47 can be attained. In particular, they proved the following result.

Theorem 2.50 [81, Theorem 3] *If G is a no-minus graph and $e \in E(G)$, then*

$$|\gamma_g(G) - \gamma_g(G-e)| \le 1 \quad and \quad |\gamma_g'(G) - \gamma_g'(G-e)| \le 1.$$

Trees are no-minus graphs, and in [81], James et al. also provided examples of trees G to demonstrate that all the claimed values of $\gamma_g(G) - \gamma_g(G-e)$ and $\gamma_g'(G) - \gamma_g'(G-e)$ in Theorem 2.50 can be achieved. Furthermore, they showed that each value in $\{0, 1, 2\}$ can be realized as $\gamma_g(T) - \gamma_g(T-v)$ and $\gamma_g'(T) - \gamma_g'(T-v)$ for some tree T and some vertex v in T, thus showing that the result of Theorem 2.49 is not changed for no-minus graphs.

The relationship between the game domination number of a graph and that of its spanning subgraphs was investigated by Brešar, Klavžar, and Rall [25]. In general, the ratio of the game domination number of a spanning subgraph to that of its supergraph can be arbitrarily large since any graph is a spanning subgraph of a complete graph of the same order. The following shows that this ratio is bounded below by $1/2$.

Theorem 2.51 [25, Proposition 4.1] *If H is a spanning subgraph of a graph G, then $\gamma_g(H) \ge \frac{\gamma_g(G)+1}{2}$.*

With ordinary domination we always have $\gamma(H) \ge \gamma(G)$ for any spanning subgraph H of a graph G. For this reason it might be somewhat surprising that the ratio $\frac{\gamma_g(H)}{\gamma_g(G)}$ can be less than 1. The following three results give a more complete picture of what can happen when comparing the game domination number of a graph and that of a spanning subgraph.

Theorem 2.52 [25, Theorem 4.2] *If $m \ge 3$, then there exists a 3-connected graph G having a 2-connected spanning subgraph H such that $\gamma_g(G) \ge 2m - 2$ and $\gamma_g(H) = m$.*

If G is a connected graph, then there is a spanning tree T of G having the same domination number. The situation can be completely different for the game domination number.

Theorem 2.53 [25, Theorem 4.4] *If k is a positive integer, then there exists a connected graph G such that $\gamma_g(T) - \gamma_g(G) \geq k$, for every spanning tree T of G.*

On the other hand, there exist graphs at the other extreme as in the following theorem.

Theorem 2.54 [25, Theorem 4.5] *If k is a positive integer, then there exists a connected graph G having a spanning tree T such that $\gamma_g(G) - \gamma_g(T) \geq k$.*

Although we do not give the explicit construction here that demonstrates Theorem 2.54, a specific example, which illustrates the same behavior for the game total domination number, is given in Section 3.4.

The behavior of $\gamma_g(G)$ and $\gamma'_g(G)$ under the contraction of an edge of G or the subdivision of an edge of G was studied in [83]. Among other results, it is proved in [83, Theorem 4] that if e is an edge of a graph G and $G.e$ the graph obtained from G by contracting the edge e, then $\gamma_g(G) - 2 \leq \gamma_g(G.e) \leq \gamma_g(G)$ and $\gamma'_g(G) - 2 \leq \gamma'_g(G.e) \leq \gamma'_g(G)$.

2.9 Domination Game (Edge-)Critical and Stable Graphs

Domination game critical graphs were introduced by Bujtás, Klavžar, and Košmrlj in [43] as follows. A graph G is *domination game critical* or simply γ_g-*critical* if $\gamma_g(G) > \gamma_g(G|v)$ holds for every $v \in V(G)$. To be more specific, we further say that if G is γ_g-critical and $\gamma_g(G) = k$, then G is k-γ_g-*critical*.

At first sight, it is not clear that the presented definition of γ_g-critical graphs is the one to be used because in other graph-theoretic concepts of criticality (like color critical graph) one considers vertex- or edge-deleted subgraphs. By Theorem 2.47 we have $\gamma_g(G - e) \in \{\gamma_g(G) - 2, \gamma_g(G) - 1, \gamma_g(G), \gamma_g(G) + 1, \gamma_g(G) + 2\}$ with all values possible. Similarly, $\gamma_g(G - v) \in \{\gamma_g(G) - 2, \gamma_g(G) - 1, \ldots\}$. So, removing an edge or a vertex can increase or decrease the game domination number. Since the Continuation Principle implies that $\gamma_g(G) \geq \gamma_g(G|v)$ holds for every $v \in V(G)$, the presented definition of domination game critical graphs is indeed the natural one.

The condition $\gamma_g(G) > \gamma_g(G|v)$ from the definition of the γ_g-critical graphs cannot be replaced with the condition $\gamma_g(G) = \gamma_g(G|v) + 1$. To see this, consider the graph G from Figure 2.4.

Using a case analysis (or computer), one can verify that G is 7-γ_g-critical. Surprisingly, $\gamma_g(G|x) = \gamma_g(G|y) = 5$ holds. But this is the largest possible surprise. More precisely, using the imagination strategy it was proved in [43, Theorem 3] that the decrease $\gamma_g(G) - \gamma_g(G|v) = 2$ is largest possible.

The only 1-γ_g-critical graph is K_1, while the 2-γ_g-critical graphs are precisely the cocktail party graphs $K_{k \times 2}$, $k \geq 1$, that is, the graphs obtained from K_{2k} by

Fig. 2.4 A 7-γ_g-critical
graph G.

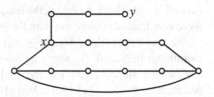

deleting a perfect matching. The variety of 3-γ_g-critical graphs is much larger as the
next result asserts.

Theorem 2.55 [43, Theorem 7] *If G is a graph with $\Delta(G) \leq n(G) - 3$, then G is
3-γ_g-critical if and only if G is twin-free and for any $v \in V(G)$ there exists a vertex
$u \in V(G)$ such that $uv \notin E(G)$ and $\deg(u) = n(G) - 3$.*

It follows from Theorem 2.55 that the join of two 3-γ_g-critical graphs is a 3-γ_g-
critical graph. The existence of k-γ_g-critical graphs for every $k \geq 1$ is guaranteed
by the following result concerning powers of cycles.

Theorem 2.56 [43, Corollary 10] *If $n \geq 1$ and $k \geq 1$, then $C^n_{2(n+1)k}$ is $(2k)$-γ_g-
critical, and $C^n_{2(n+1)k+1}$ is $(2k+1)$-γ_g-critical.*

There are no γ_g-critical trees up to 12 vertices. After that, there are two γ_g-critical
trees on 13 vertices, no such trees on 14 or 15 vertices, one more on 16 vertices, but
ten on 17 vertices. In summary, up to order 17, there are only 13 γ_g-critical trees.
In [52] Dorbec, Henning, Klavžar, and Košmrlj extended the list up to order 20; the
number of γ_g-critical trees on 18, 19, and 20 vertices is 14, 18, and 13, respectively.

Let $T_{p,q,r}$ be the tree obtained from disjoint paths P_{4p+1}, P_{4q+1}, and P_{4r+1} by
identifying three end-vertices, one from each path. The following result verifies the
corresponding conjecture stated in [43, p. 792].

Theorem 2.57 [52, Theorem 4.1] *If $p, q, r \geq 1$, then $T_{p,q,r}$ is a $(2(p+q+r)+1)$-
γ_g-critical graph.*

To be able to prove Theorem 2.57, new tools were developed in [52], the so-
called Cutting Lemma and the Union Lemma. Since these lemmas may prove useful
in their own right, we present their statements. First we introduce some additional
notation. Let G be a graph and uv an arbitrary edge in G. Let G_{uv} be the graph
obtained from $G - uv$ by adding two new vertices u' and v', adding the two edges
uv' and vu', and declaring that both u' and v' are dominated.

Lemma 2.58 (Cutting Lemma) [52, Theorem 3.1] *Let G be a graph, and let
$A, B \subseteq V(G)$ where $B \subseteq A$. If uv is an edge of G, then*

$$\gamma_g(G|A) \leq \gamma_g(G_{uv}|B) \quad and \quad \gamma'_g(G|A) \leq \gamma'_g(G_{uv}|B).$$

Proof. We proceed by induction on $|V(G)-A|$. If $|V(G)-A| = 0$, then $\gamma_g(G|A) =
0$ and $\gamma_g(G_{uv}|B) \geq 0$, and $\gamma'_g(G|A) = 0$ and $\gamma'_g(G_{uv}|B) \geq 0$, thereby proving the

desired result. This establishes the base case. Suppose that $|V(G) - A| \geq 1$. Hence, there is at least one more move in the game on $G|A$.

We first prove that $\gamma_g(G|A) \leq \gamma_g(G_{uv}|B)$. Let x be an optimal first move for Dominator in $G_{uv}|B$. By applying the Continuation Principle, we may assume that $x \notin \{u', v'\}$, since $N_{G_{uv}}[u'] \subseteq N_{G_{uv}}[v]$ and $N_{G_{uv}}[v'] \subseteq N_{G_{uv}}[u]$. Since $B \subseteq A$, we deduce that $B \cup N_{G_{uv}}[x] \subseteq A \cup N_G[x] \cup \{u', v'\}$. We now consider two possibilities.

Suppose, firstly, that x is a legal move in $G|A$. In this case, we have $N_G[x] - A \neq \emptyset$. We infer

$$\gamma_g(G_{uv}|B) = 1 + \gamma_g' \left(G_{uv}|(B \cup N_{G_{uv}}[x]) \right)$$
$$\geq 1 + \gamma_g'(G|(A \cup N_G[x]))$$
$$\geq \gamma_g(G|A)$$

where the first inequality applies induction, while the second inequality follows from the fact that x may not be an optimal first move of Dominator in $G|A$.

Suppose, secondly, that x is not a legal move in $G|A$. In this case, Dominator may choose any other move y and we still have $B \cup N_{G_{uv}}[x] \subseteq A \subseteq A \cup N_G[y] \cup \{u', v'\}$. We again apply induction in the first inequality and then use the fact that y may not be an optimal first move of Dominator in $G|A$:

$$\gamma_g(G_{uv}|B) = 1 + \gamma_g'(G_{uv}|(B \cup N_{G_{uv}}[x]))$$
$$\geq 1 + \gamma_g'(G|(A \cup N_G[y]))$$
$$\geq \gamma_g(G|A).$$

We prove next that $\gamma_g'(G|A) \leq \gamma_g'(G_{uv}|B)$. Let x be an optimal first move for Staller in $G|A$. We note that $N_{G_{uv}}[x] - \{u', v'\} \subseteq N_G[x]$. Further we note that since $B \subseteq A$, we have $B \cup N_{G_{uv}}[x] \subseteq A \cup N_G[x] \cup \{u', v'\}$. We now consider two possibilities.

Suppose that $N_{G_{uv}}[x] - B$ is not a subset of $\{u', v'\}$. In this case,

$$\gamma_g'(G|A) = 1 + \gamma_g(G|(A \cup N_G[x])) \qquad \text{(x is optimal for Staller)}$$
$$\leq 1 + \gamma_g(G_{uv}|(B \cup N_{G_{uv}}[x])) \qquad \text{(by induction)}$$
$$\leq \gamma_g'(G_{uv}|B)$$

since x is not necessarily an optimal move for Staller in $G_{uv}|B$.

Suppose next that $N_{G_{uv}}[x] - B \subset \{u', v'\}$. Therefore, x is u or v and the only newly dominated vertex in $G|A$ is v or u. Renaming vertices if necessary, we may assume that $x = u$. Thus, u' is a legal move in $G_{uv}|B$ that newly dominates exactly the same vertex, that is, $B \cup N_{G_{uv}}[u'] \subseteq A \cup N_G[x] \cup \{u', v'\}$. We now apply induction with

$$\gamma_g'(G|A) = 1 + \gamma_g(G|(A \cup N_G[x])) \qquad \text{(x is optimal for Staller)}$$
$$\leq 1 + \gamma_g(G_{uv}|(B \cup N_{G_{uv}}[u'])) \qquad \text{(by induction)}$$
$$\leq \gamma_g'(G_{uv}|B)$$

noting that u' may not be an optimal move for Staller in $G_{uv}|B$. This completes the proof of Lemma 2.58.   □

As remarked in [52], the above proof follows the idea of the imagination strategy. Indeed in the first half of the proof, $G_{uv}|B$ plays the role of the imagined game for Dominator, and in the second half of the proof, $G|A$ does so for Staller.

In order to state the Union Lemma, we define the *weighting function* w of paths as follows (by P'_n we denote $P_n|u$ and by P''_n we denote $P_n|\{u, v\}$, where u and v are the two leaves of P_n):

$$w(P'_{4q+r}) = w(P''_{4q+r}) = 2q + \begin{cases} 0; & r = 0, \\ 1; & r = 1, \\ \frac{3}{2}; & r = 2, \\ \frac{7}{4}; & r = 3. \end{cases}$$

We are now in a position to state the Union Lemma.

Lemma 2.59 (Union Lemma) [52, Lemma 3.2] *If $F_1, \ldots, F_k$ are vertex-disjoint paths where $F_i = P'_{n_i}$ or $F_i = P''_{n_i}$ for $i \in [k]$ and $n_i \geq 1$, then*

$$\gamma'_g\left(\bigcup_{i=1}^{k} F_i\right) \leq \left\lceil \sum_{i=1}^{k} w(F_i) \right\rceil.$$

Domination game stable graphs were studied by Xu and Li [111]. A graph G is *domination game stable* or γ_g-*stable* if $\gamma_g(G) = \gamma_g(G|v)$ holds for every vertex $v \in V(G)$. We further say that if G is γ_g-stable and $\gamma_g(G) = k$, then G is k-γ_g-*stable*.

It is easy to see that connected 1-γ_g-stable graphs are precisely those graphs that have a dominating vertex [111, Proposition 2.1]. In addition, connected graphs that are 2-γ_g-stable were characterized [111, Theorem 2.3], and a structural characterization of 3-γ_g-stable trees was presented by Xu and Li [111, Theorem 2.5]. It is worth mentioning that among 3-γ_g-stable trees, there are only caterpillars. As a sporadic family of 3-γ_g-stable graphs, one can find Kneser graphs $K(n, 2)$, where $n \geq 6$ [111, Proposition 2.12]. Interestingly, the Petersen graph $K(n, 2)$ is 5-γ_g-stable [111, Proposition 2.11]. The paths P_n, where $n \geq 2$, and the cycles C_n, where $n \geq 3$, are γ_g-stable if and only if $n \equiv 2 \pmod 4$ or $n \equiv 3 \pmod 4$; see [111, Theorems 2.8 and 2.10]. We end this brief presentation of what is known about γ_g-stable graphs with the following result.

Theorem 2.60 [111, Theorem 2.13] *For an arbitrary positive integer t, there exists a t-γ_g-stable graph.*

For an odd integer t, $t > 3$, a family of t-γ_g-stable graphs is presented in Figure 2.5. The graph X_k is obtained from C_6 by attaching to it a pendant vertex x, and then attaching s, $s \geq 2$, leaves to the vertex x, and also attaching k,

Fig. 2.5 The graph X_k with
$k \geq 1$ and $s \geq 2$.

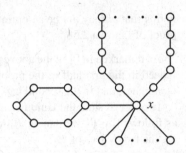

$k \geq 1$, disjoint paths of length 5 to the vertex x. It is proved in [111] that
$\gamma_g(X_k) = 2k + 3 = \gamma_g(X_k|v)$, where v is any vertex of X_k. For an even integer
t, a similar construction works, only that the cycle C_6 in X_k is replaced by the cycle
C_3. The resulting graph Y_k is $(2k + 2)$-γ_g-stable, which covers all even integers t,
$t > 2$.

Xu and Li considered also edge-critical graphs for the game domination number [111]. It was mentioned in Section 2.8 that given a graph G and $e \in E(G)$, the difference $\gamma_g(G) - \gamma_g(G - e)$ can achieve any value from $\{-2, -1, 0, 1, 2\}$. A graph G is *domination game edge-critical*, or γ_g-*edge-critical* for short, if for every edge e,

$$\gamma_g(G) - \gamma_g(G - e) \in \{-1, -2\}.$$

That is, G is γ_g-edge-critical if the game domination number increases when an arbitrary edge is removed from G. We further say that if G is γ_g-edge-critical and $\gamma_g(G) = k$, then G is k-γ_g-*edge-critical*.

Domination game edge-critical graphs G with $\gamma_g(G) = 1$ are exactly the stars $K_{1,n}$ [111, Proposition 3.2]. Furthermore, a characterization of 2-γ_g-edge-critical graphs is given in [111, Theorem 3.3], and an example, which is at the same time 3-γ_g-edge-critical and 3-γ_g-stable, is presented [111]. Several sufficient conditions for a graph not to be γ_g-edge-critical are also proved in [111]. In particular, if v is an optimal first move of Dominator in a D-game, and v has a neighbor u with $\deg(u) = 2$, then G is not γ_g-edge-critical [111, Theorem 3.8]. This implies that the paths P_n and the cycles C_n are not γ_g-edge-critical when $n \geq 4$ [111, Corollary 3.9].

It seems that finding examples of γ_g-edge-critical graphs is even more difficult than finding γ_g-stable graphs. Since all examples of γ_g-edge-critical graphs found by Xu and Li are also γ_g-stable, they asked whether every k-γ_g-edge-critical graph is also k-γ_g-stable [111, Problem 4.3]. The question has a positive answer if $k \in \{1, 2\}$. Finally, it is also open whether there exists a tree different from a star, which is γ_g-edge-critical [111, Problem 4.4].

2.10 Perfect Graphs for the Domination Game

Recall from Section 2.5 that a graph G is a γ_g-minimal if $\gamma_g(G) = \gamma(G)$ holds, and that G is γ_g'-minimal if $\gamma_g'(G) = \gamma(G)$ holds. Bujtás, Iršič, and Klavžar [37] introduced the hereditary versions of these properties as follows. A graph G is called γ_g-*perfect* if every induced subgraph F of G satisfies $\gamma_g(F) = \gamma(F)$, and G is γ_g'-*perfect*, if every induced subgraph F of G satisfies $\gamma_g'(F) = \gamma(F)$.

In the main result from [37], γ_g-perfect graphs are characterized. To state the theorem, some preparation is needed. First, a graph G is 2-γ_g-*perfect*, if every induced subgraph F of G with $\gamma(F) = 2$ is a γ_g-minimal graph. Second, we need another operation defined as follows.

A *perfect set of cliques* (PSC) in a graph G is a (possibly empty) set Q of homogeneous cliques (that is, every two vertices in each of the cliques are twins) such that $d_G(Q, Q') = 3$ and there is a join between $N_G[Q] - Q$ and $N_G[Q'] - Q'$, for every pair Q and Q' of distinct cliques in Q. Now, if G is a graph, v a new vertex, and Q a perfect set of cliques in G, then the graph $O(G, v, Q)$ is obtained from G by adding the vertex v and making it adjacent to all vertices in $V(G) - V(Q)$. We can now state:

Theorem 2.61 [37, Theorem 3.7] *If G is a graph, then the following statements are equivalent:*

(i) *G is γ_g-perfect.*
(ii) *G is 2-γ_g-perfect.*
(iii) *G can be obtained from an isolated vertex by repeatedly applying the following operators:*

  – *For a graph F, and for an $s \in \mathbb{N}$, take $F \sqcup K_s$.*
  – *For a graph F, and for a PSC Q in F, take $O(F, v, Q)$.*

The construction from Theorem 2.61(iii) is illustrated in Figure 2.6. The example indicates that the variety of γ_g-perfect graphs is large.

An important consequence of Theorem 2.61 is that γ_g-perfect graphs can be recognized in polynomial time [37, Theorem 4.1].

The proof of Theorem 2.61 is quite involved, a key idea of it is that to determine whether a graph is γ_g-perfect it suffices to consider its induced subgraphs with

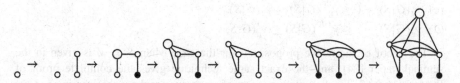

Fig. 2.6 A construction of γ_g-perfect graphs. At the steps where operator O is applied, the vertices from the perfect set of cliques are black and the newly added vertex is larger.

domination number 2. On the other hand, it is straightforward to characterize γ_g'-perfect graphs, they are precisely disjoint unions of cliques [37, Proposition 5.1].

In [37], *minimally γ_g-imperfect graphs* were also introduced as the graphs G such that each proper induced subgraph of G is γ_g-perfect, but $\gamma(G) < \gamma_g(G)$. It was proved that the only disconnected, minimally γ_g-imperfect graph is $2P_3$ [37, Theorem 4.4]. A list of minimally γ_g-imperfect triangle-free graphs was also determined [37, Theorem 4.6]. In addition, the following was proved:

Theorem 2.62 [37, Theorem 3.8] *Every minimally γ_g-imperfect graph has domination number 2.*

2.11   PLUS, MINUS, and Bluff Graphs

Recall that a graph G realizes the pair (k, ℓ) if $\gamma_g(G) = k$ and $\gamma_g'(G) = \ell$. By Theorem 2.2 for a given k, we have $\ell \in \{k-1, k, k+1\}$. A graph that realizes $(k, k+1)$, (k, k), and $(k, k-1)$ is called a PLUS graph, an EQUAL graph, or a MINUS graph, respectively.

Dorbec, Košmrlj, and Renault [53] define a *no-minus graph* as a graph G such that for every subset $S \subseteq V(G)$, $\gamma_g(G|S) \le \gamma_g'(G|S)$; in other words, $G|S$ is not a MINUS graph. By Theorem 2.3, every partially dominated forest with no isolated vertex is a no-minus graph. Next we present properties of no-minus graphs.

First we define two variants of the game (both D-game and S-game) where one of the players is allowed to pass a move and this is not counted as a move in the game domination number. Let $\gamma_g^{dp}(G)$ and $\gamma_g'^{dp}(G)$ be the number of moves in a D-game and S-game, respectively, when both players play optimally and Dominator is allowed to pass at most one move (at any time in the game). Let $\gamma_g^{sp}(G)$ and $\gamma_g'^{sp}(G)$ be the number of moves in a D-game and S-game, respectively, when both players play optimally and Staller is allowed to pass at most one move (at any time in the game). Dorbec et al. observed the following properties of the variants of the games where one player is permitted to pass a move.

Proposition 2.63 [53, Lemma 2.2] *If G is a graph and $S \subseteq V(G)$, then*

(a) $\gamma_g(G|S) \le \gamma_g^{sp}(G|S) \le \gamma_g(G|S) + 1$,
(b) $\gamma_g'(G|S) \le \gamma_g'^{sp}(G|S) \le \gamma_g'(G|S) + 1$,
(c) $\gamma_g(G|S) - 1 \le \gamma_g^{dp}(G|S) \le \gamma_g(G|S)$,
(d) $\gamma_g'(G|S) - 1 \le \gamma_g'^{dp}(G|S) \le \gamma_g'(G|S)$.

The proof of the above proposition for the case when $S = \emptyset$ is given in the seminal paper [24], and the exact same technique gives the complete proof of Proposition 2.63. The first nice property of no-minus graphs is that passing a move has no advantage for either player.

Proposition 2.64 [53, Proposition 2.3] *If G is a no-minus graph and $S \subseteq V(G)$, then*

(a) $\gamma_g^{sp}(G|S) = \gamma_g^{dp}(G|S) = \gamma_g(G|S)$,

(b) $\gamma_g^{'sp}(G|S) = \gamma_g^{'dp}(G|S) = \gamma_g'(G|S)$.

The next useful property of no-minus graphs considers the game domination numbers of disjoint unions of graphs.

Theorem 2.65 [53, Theorem 2.12] *If $G_1|S_1$ and $G_2|S_2$ are partially dominated no-minus graphs such that $G_1|S_1$ realizes $(k, k + 1)$ and $G_2|S_2$ realizes $(\ell, \ell + 1)$, then*

(a) $k + \ell \leq \gamma_g((G_1 \sqcup G_2)|(S_1 \cup S_2)) \leq k + \ell + 1$,

(b) $k + \ell + 1 \leq \gamma_g'((G_1 \sqcup G_2)|(S_1 \cup S_2)) \leq k + \ell + 2$.

Beside forests, there are other families that enjoy the useful property of being no-minus graphs. As shown by Dorbec, Košmrlj, and Renault [53], the so-called tri-split graphs and the well known dually-chordal graphs are also no-minus graphs.

In general, we have the following bounds on the game domination number of disjoint unions of graphs.

Theorem 2.66 [53, Corollary 3.2] *If $G_1|S_1$ and $G_2|S_2$ are two partially dominated graphs, then*

(a) $\gamma_g((G_1 \sqcup G_2)|(S_1 \cup S_2)) \geq \gamma_g(G_1|S_1) + \gamma_g(G_2|S_2) - 1$,

(b) $\gamma_g((G_1 \sqcup G_2)|(S_1 \cup S_2)) \leq \gamma_g(G_1|S_1) + \gamma_g'(G_2|S_2) + 1$,

(c) $\gamma_g'((G_1 \sqcup G_2)|(S_1 \cup S_2)) \leq \gamma_g'(G_1|S_1) + \gamma_g'(G_2|S_2) + 1$,

(d) $\gamma_g'((G_1 \sqcup G_2)|(S_1 \cup S_2)) \geq \gamma_g'(G_1|S_1) + \gamma_g(G_2|S_2) - 1$.

2.11.1 Bluff Graphs

In general, it is extremely difficult to find optimal first moves on an arbitrary graph for each of the players in the domination game. In some cases, the situation can be just the opposite. For instance, it is clear that if G is vertex-transitive, then the first player to select a vertex can do this task arbitrarily. In order to increase the variety of such graphs, *bluff graphs* were introduced by Brešar, Dorbec, Klavžar, and Košmrlj, in [15] as the graphs in which every vertex is an optimal first move for Dominator in the D-game, and every vertex is also an optimal first move for Staller in the S-game.

Theorem 2.67 [15, Theorem 2.2] *Every MINUS graph is a bluff graph. Moreover, if G is a connected graph with $\gamma_g(G) \geq 2$ and $\delta(G) = 1$, then G is a bluff graph if and only if G is a MINUS graph.*

Proof. Suppose G realizes $(k, k - 1)$, and let u be an arbitrary vertex of G. Suppose that in the D-game Dominator plays u as his first move. Then by Corollary 2.4, $\gamma_g'(G|N[u]) = k - 1$, which implies that u is an optimal first move for Dominator.

By the same corollary, after Staller plays u as her first move in the S-game, $\gamma_g(G|N[u]) = k - 2$ holds. Hence u is also an optimal first move for Staller in the S-game. It follows that G is a bluff graph.

For the second assertion assume that G is a connected, bluff graph with $\gamma_g(G) \geq 2$, and let u be a leaf of G. Let w be the support vertex of u. Since G is a bluff graph, $d_1 = u$ is an optimal first move in the D-game. Suppose that $s_1 = w$. This is a legal move because $\gamma_g(G) \geq 2$ and because G is connected. The move of Staller may not be optimal, hence $\gamma_g(G|N[w]) \leq \gamma_g(G) - 2$. On the other hand, since G is a bluff graph, the move $s_1' = w$ in the S-game implies $\gamma_g(G|N[w]) = \gamma_g'(G) - 1$. It follows that $\gamma_g'(G) - 1 \leq \gamma_g(G) - 2$, that is, $\gamma_g'(G) \leq \gamma_g(G) - 1$. By Theorem 2.2 it follows that $\gamma_g'(G) = \gamma_g(G) - 1$, that is, G is a MINUS graph.   □

2.12 Computational Complexity

The algorithmic complexity of determining the game domination number of a given graph was studied by Brešar, Dorbec, Klavžar, Košmrlj, and Renault [16], who show that the complexity of verifying whether the game domination number of a graph is bounded by a given integer is in the class of PSPACE-complete problems. Formally, they consider the following two game domination problems.

D-GAME DOMINATION PROBLEM

Input: A graph G, and an integer ℓ.
Question: Is $\gamma_g(G) \leq \ell$?

S-GAME DOMINATION PROBLEM

Input: A graph G, and an integer ℓ.
Question: Is $\gamma_g'(G) \leq \ell$?

They then present a reduction to the Game Domination Problem from the POS-CNF problem, which is known to be log-complete in PSPACE (see [105] for the complexity result on this problem) as follows. In POS-CNF we are given a set of variables, and a formula that is a conjunction of disjunctive clauses, in which there are no negations of variables. Two players alternate turns, the first player setting a previously unset variable TRUE, and the second player setting such a variable FALSE. After all variables are set, the first player wins if the formula is TRUE, otherwise the second player wins.

A given formula $\mathcal{F}$ using k variables and n disjunctive clauses is transformed into a partially dominated graph $G_{\mathcal{F}}|A$, having $9k + n + 4$ vertices. For this purpose, the authors used a partially dominated gadget graph $W|\{a_1, a_2\}$ illustrated in Figure 2.7,

Fig. 2.7 The gadget graph $W|\{a_1, a_2\}$ representing each variable.

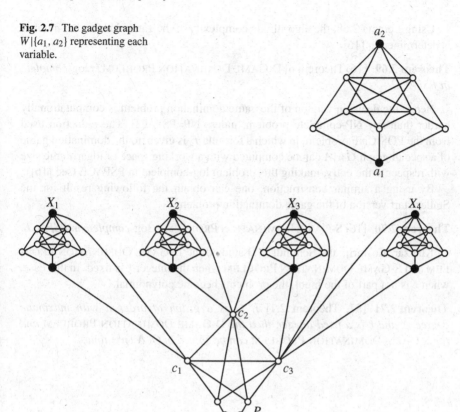

Fig. 2.8 Example of the graph for formula $X_1 \wedge (X_2 \vee X_3) \wedge (X_3 \vee X_4)$.

where the vertices a_1 and a_2 are marked as vertices that are already dominated. Each copy of $W|\{a_1, a_2\}$ corresponds to a variable of the POS-CNF problem.

As explained in [16], given a formula $\mathcal{F}$ with k variables and n clauses, a construction of the graph $G_{\mathcal{F}}|A$ is described as follows. For each variable X in $\mathcal{F}$, a copy W_X of the graph $W|\{a_1, a_2\}$ is taken. For each disjunctive clause C_i in the formula a vertex c_i is added, and for each X that appears in C_i the vertex c_i is joined to both a_1 and a_2 from the copy of W_X. Next, edges $c_i c_j$ are added between each two vertices, corresponding to disjunctive clauses C_i, C_j that appear in $\mathcal{F}$, so that the vertices c_i, $i \in [n]$ induce a clique Q of size n. Finally, a copy $P : p_1 p_2 p_3 p_4$ of the path P_4 is added, and the edges $p_1 c_i$ and $p_4 c_i$ are added for $i \in [n]$. See Figure 2.8 for an example of the construction. (Note that A is the union of the sets $\{a_1, a_2\}$ over all copies W_X.)

Lemma 2.68 [16, Corollary 5] *Player 1 has a winning strategy for a formula $\mathcal{F}$ in the POS-CNF game if and only if $\gamma_g(G_{\mathcal{F}}|A) \leq 3k + 2$.*

Using Lemma 2.68, the algorithmic complexity of the game domination number is determined in [16].

Theorem 2.69 [16, Theorem 6] D-GAME DOMINATION PROBLEM *is* log-*complete in PSPACE.*

Hence the decision version of the game domination problem is computationally harder than any NP-complete problem, unless NP=PSPACE. The reduction used from the POS-CNF problem, in which a formula $\mathcal{F}$ is given, to the domination game of a special graph $G_{\mathcal{F}}|A$ can be computed with a working space of logarithmic size with respect to the entry, making this problem log-complete in PSPACE (see [16]).

By using a similar construction, one can obtain the following result for the Staller-start version of the game domination problem.

Theorem 2.70 [16] S-GAME DOMINATION PROBLEM *is* log-*complete in PSPACE.*

Klavžar, Košmrlj, and Schmidt [87] studied the D-GAME DOMINATION PROBLEM and S-GAME DOMINATION PROBLEM when the integer ℓ is fixed. In this case, when ℓ is not part of the input, the problems become polynomial.

Theorem 2.71 [87, Theorem 2.1] *If G is a graph of order n with maximum degree Δ and ℓ is a fixed integer, then the* D-GAME DOMINATION PROBLEM *and the* S-GAME DOMINATION PROBLEM *can be solved in* $O(\Delta \cdot n^{\ell})$ *time.*

Chapter 3
Total Domination Game

Total domination is the second most studied topic in domination theory, and thus the total domination game is a natural variation of the domination game. It was introduced and first studied in 2015 by Henning, Klavžar, and Rall [70]. There are, of course, some similarities between these two kinds of domination games, but it was shown in this introductory paper that these versions differ significantly. As mentioned in Chapter 1, the total domination game is well defined just on isolate-free graphs. Thus, in this chapter we restrict our discussion to such graphs, and even if we do not emphasize it, the considered graphs have no isolated vertices.

The structure of this chapter is in many ways parallel to Chapter 2. It begins in Section 3.1 with the important and often-used Total Continuation Principle, whose proof relies on the imagination strategy and is analogous to that of the Continuation Principle in Section 2.1. This is followed by two sections that discuss upper bounds for the game total domination number for graphs of order n, all of whose components have order at least 3. The conjectured upper bound is $\frac{3n}{4}$, and Section 3.3 presents what is known about this conjecture. Then the presentation turns to the total domination game on trees. A characterization of the trees with equal total domination number and game total domination number is given, and it is shown that there are graphs whose game total domination number is much larger than that of one of its spanning trees. Similar to the domination game, it is rare to be able to find and verify the game total domination number for an arbitrary graph. In Section 3.5 formulas are given for the game total domination number of paths and cycles, and an infinite family is presented whose graphs G satisfy $\gamma_{tg}(G) = 2\gamma_t(G) - 1$, the general upper bound. The effect on the game total domination number of a single vertex being removed or of being totally dominated (before the game begins) is studied in the next section, and we close out the chapter with a discussion of the computational complexity of the game total domination number.

B. Brešar et al., *Domination Games Played on Graphs*, SpringerBriefs in Mathematics, https://doi.org/10.1007/978-3-030-69087-8_3

3.1   Total Continuation Principle

As remarked in Section 2.1, the proof of the Continuation Principle presented in
Lemma 2.1 could be modified to work on some variants of the domination game.
Indeed, the Continuation Principle also holds for the total domination game as first
shown in [70, Lemma 2.1]. Similarly as for the standard domination game, we define
a *partially total dominated graph* $G|S$ as a graph G together with a declaration that
vertices of the set $S \subseteq V(G)$ are already totally dominated and need not be totally
dominated in the rest of the game.

Lemma 3.1 (Total Continuation Principle) *If G is a graph and $A, B \subseteq V(G)$ with
$B \subseteq A$, then $\gamma_{tg}(G|A) \le \gamma_{tg}(G|B)$ and $\gamma'_{tg}(G|A) \le \gamma'_{tg}(G|B)$.*

As a consequence of the Total Continuation Principle, whenever x and y are
legal moves for Dominator in the total domination game and $N(x) \subseteq N(y)$, we
may assume that Dominator will play y instead of x. Similarly, if it was Staller's
turn, we may assume that Staller will play x instead of y. As with the domination
game, the Total Continuation Principle implies that in the total domination game the
number of moves in the D-game and the S-game when played optimally can differ
by at most 1.

Theorem 3.2 [70, Theorem 2.2] *If G is a graph, then $|\gamma_{tg}(G) - \gamma'_{tg}(G)| \le 1$.*

As observed in [70], there are graphs G_1, G_2, and G_3 such that $\gamma_{tg}(G_1) =
\gamma'_{tg}(G_1)$, $\gamma_{tg}(G_2) = \gamma'_{tg}(G_2) + 1$, and $\gamma_{tg}(G_3) = \gamma'_{tg}(G_3) - 1$. For example,
$\gamma_{tg}(P_4) = \gamma'_{tg}(P_4) = 3$, $\gamma_{tg}(P_5) = 3 = \gamma'_{tg}(P_5) - 1$, and $\gamma_{tg}(C_8) = 5 = \gamma'_{tg}(C_8) + 1$.
Infinite families of such examples were obtained in [70].

Using the Total Continuation Principle, Henning and Rall [76] show that the
number of moves in the D-game in a partially total dominated forest with no isolated
vertex can never exceed the number of moves in the S-game.

Theorem 3.3 [76, Theorem 1] *If F is a partially total dominated forest with no
isolated vertex, then $\gamma_{tg}(F) \le \gamma'_{tg}(F)$.*

3.2   Upper Bounds for General Graphs

If a graph G is the disjoint union of copies of K_2, then $\gamma_{tg}(G) = n(G)$. Hence it
is only of interest to study upper bounds on the game total domination number of a
graph in which every component contains at least three vertices.

The first such general upper bound was obtained in 2016 by Henning, Klavžar,
and Rall in [71]. Their proof adopts Bujtás's coloring approach for the ordinary
domination game, except here they color the vertices of a graph with four colors that
reflect four different types of vertices. Weights are then assigned to colored vertices
and the decrease of the total weight of the graph as a consequence of playing vertices
in the course of the game is analyzed according to three possible phases of the game.

In each phase of the game, a strategy of Dominator is described and analyzed that will guarantee that the game is complete after the number of moves is at most four-fifths the order of the graph.

Theorem 3.4 [71, Theorem 1] *If G is a graph of order n in which every component contains at least three vertices, then* $\gamma_{tg}(G) \leq \frac{4}{5}n$ *and* $\gamma'_{tg}(G) \leq \frac{4n+2}{5}$.

In 2018 Bujtás [31] obtained a new general upper bound on the game total domination number that improves this earlier bound of $\frac{4}{5}n$. We remark that in order to prove this result, Bujtás [31] established an upper bound on the game transversal number of a hypergraph in which edges of size one are not excluded. The transversal game we discuss in Section 5.5. Her ingenious proof defines the so-called special vertices and edges in the hypergraph, and thereafter proceeds to assign weights to the vertices and edges, both special and non-special. The game is then studied according to several possible phases of the game.

Theorem 3.5 [31, Theorem 2 and Proposition 15] *If G is a graph of order n in which every component contains at least three vertices, then* $\gamma_{tg}(G) \leq \frac{11}{14}n$ *and* $\gamma'_{tg}(G) \leq \frac{11n+6}{14}$.

3.3 $\frac{3}{4}$-Conjecture

Bujtás's bound in Theorem 3.5 is the best general upper bound on the game total domination number to date. Much of the recent interest in the total domination game arose from attempts to determine the best possible upper bound on the game total domination number. In 2016, Henning, Klavžar, and Rall [71] posed the total domination game $\frac{3}{4}$-**Conjecture**.

Conjecture 3.6 [71] If G is a graph in which every component contains at least three vertices, then $\gamma_{tg}(G) \leq \frac{3}{4}n(G)$.

We discuss next special cases where the total domination game $\frac{3}{4}$-Conjecture has been proven. In 2016, Bujtás, Henning, and Tuza [34] proved that the $\frac{3}{4}$-Conjecture holds over the class of graphs with minimum degree at least 2. To do this, they raise the problem to a higher level by introducing a transversal game in hypergraphs, and establish a tight upper bound on the game transversal number of a hypergraph with all edges of size at least 2 in terms of its order and size. Their exact result we present in Section 5.5 where we discuss the transversal game in hypergraphs in more detail. We remark here that the result of Theorem 5.35 presented in Section 5.5 validates, as a by-product, the total domination game $\frac{3}{4}$-Conjecture on graphs with minimum degree at least 2 noting that $\frac{8}{11} < \frac{3}{4}$.

Corollary 3.7 [34, Corollary 2] *The total domination game $\frac{3}{4}$-Conjecture is true over the class of graphs with minimum degree at least 2.*

Subsequently, in 2016 Henning and Rall [75] proved that the total domination
game $\frac{3}{4}$-Conjecture holds in a general graph G (with no isolated vertex) if we
remove the minimum degree at least 2 condition, but impose the weaker condition
that the degree sum of adjacent vertices in G is at least 4 and add the requirement
that no two leaves are at distance 4 apart in G. We state their result formally as
follows.

Theorem 3.8 [75, Theorem 2] *The total domination game $\frac{3}{4}$-Conjecture is true
over the class of graphs G that satisfy both conditions* (a) *and* (b) *below:*

(a) *The degree sum of adjacent vertices in G is at least* 4.
(b) *No two leaves are at distance exactly* 4 *in G.*

As a special case of Theorem 3.8, we recover the result of Corollary 3.7 that
the total domination game $\frac{3}{4}$-Conjecture is valid on graphs with minimum degree
at least 2. We remark that the approach in [75] is to color the vertices with six
colors that reflect six different types of vertices and to associate a weight with each
vertex. They then study the decrease of total weight of the graph as a consequence
of playing vertices in the course of the game. Although the result of Theorem 5.35
is a stronger result than Corollary 3.7, the proof of Corollary 3.7 given in [75] is
surprising in that Dominator can complete the total domination game played in a
graph with minimum degree at least 2 in at most $3n/4$ moves by simply following
a greedy strategy in the associated colored-graph, where a greedy strategy plays a
move that decreases the total weight of the graph as much as possible.

Despite the excellent progress made over the past few years, the total domination
game $\frac{3}{4}$-Conjecture has yet to be settled.

As remarked in [71], if the total domination game $\frac{3}{4}$-Conjecture is true, then the
upper bound is best possible. The simplest examples of graphs achieving equality
in the conjectured bound are the path P_4 and the path P_8, noting that $\gamma_{tg}(P_4) =
\gamma'_{tg}(P_4) = 3$ and $\gamma_{tg}(P_8) = \gamma'_{tg}(P_8) = 6$.

We note that by taking, for example, $G \cong kP_8$ where $k \geq 1$, the optimal strategy
of Staller is whenever Dominator starts playing on a component of G to play on
that component and adopt her optimal strategy on the component. Since $\gamma_{tg}(P_8) =
6$, which is even, she can continue this strategy until the completion of the game.
This shows that $\gamma_{tg}(G) = 6k = \frac{3}{4}n$, where $n = 8k$ is the number of vertices in G.

We show next the existence of a tree of order 12 whose game total domination
number is exactly three-quarters its order. Let T_{12} be the tree obtained from two
vertex disjoint paths of order 6 by adding an edge between a vertex at distance 2
from a leaf in the one path to a vertex at distance 2 from a leaf in the other path. The
tree T_{12} of order $n = 12$ is illustrated in Figure 3.1.

Fig. 3.1 A tree T_{12} of order n
satisfying $\gamma_{tg}(T_{12}) = \frac{3}{4}n$.

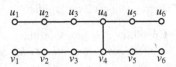

A computer check verifies that every move for Dominator is optimal as his first move in the total domination game played in T_{12}, except for the vertices u_6 and v_6. We illustrate here one possible sequence of optimal moves for the total domination game played in T_{12}. Suppose that Dominator plays the vertex $d_1 = v_4$ as his first move. In this case, the vertex $s_1 = u_6$ is the only optimal move for Staller. Every legal move for Dominator is optimal on his second move. If Dominator now plays the vertex $d_2 = v_3$, for example, as his second move, then the vertex $s_2 = u_4$ is the only optimal move for Staller. Thereafter, all four support vertices, namely u_2, u_5, v_2, and v_5, must be played, in addition to exactly one of u_1 and u_3. Thus in this case, nine moves are made to complete the game. With a more extensive analysis (or using a computer), it can be shown that $\gamma_{tg}(T_{12}) = 9 = \frac{3}{4}n$. Moreover, $\gamma'_{tg}(T_{12}) = 9$. We state the above observations formally as follows.

Proposition 3.9 *If* $T \in \{P_4, P_8, T_{12}\}$, *then* $\gamma_{tg}(T) = \gamma'_{tg}(T) = \frac{3}{4}n(T)$.

We pose the following problem that has yet to be settled.

Problem 3.10 Determine the connected graphs G satisfying $\gamma_{tg}(G) = \frac{3}{4}n(G)$.

3.4 Total Domination Game on Trees

Henning and Rall [76] presented a characterization of trees with equal total domination and game total domination numbers. For this purpose, they constructed a family $\mathcal{F}^*$ of trees as follows. Let x be a specified vertex in a tree T and define the following types of attachments at the vertex x that are used to build larger trees. In all cases, we call the vertex of the attachment that is joined to x the *link vertex* of the attachment.

- For $i \in [3]$, an *attachment of Type-i at x* is an operation that adds a path P_{i+1} to T and joins one of its ends to x.
- An *attachment of Type-A at x* is obtained by adding an attachment of Type-1 at x with link vertex x', followed by at least one attachment of Type-2 at x'.
- An *attachment of Type-B at x* is obtained by adding an attachment of Type-A at x, followed by an attachment of Type-3 to at least one new (added) vertex at distance 3 from x.

We note that each attachment of Type-A at x can be obtained from a star $K_{1,k}$, for some $k \geq 2$, by subdividing $k - 1$ edges twice and joining the central vertex of the original star to x.

For integers $k_1, k_2, k_3, k_4 \geq 0$, let $\mathcal{T}_{k_1,k_2,k_3,k_4}$ be the family of all trees obtained from a trivial tree K_1 whose vertex is named a by applying k_i attachments of Type-i at a for each $i \in [2]$, applying k_3 attachments of Type-A at a and applying k_4 attachments of Type-B at a. Let

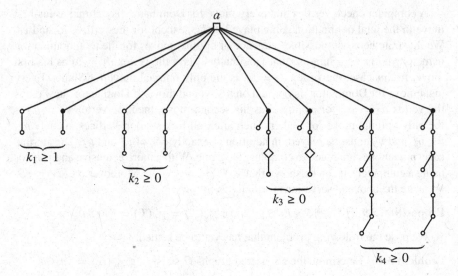

Fig. 3.2 A tree in the family $\mathcal{T}_{k_1,k_2,k_3,k_4}$.

Fig. 3.3 The tree F_{10}.

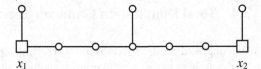

$$\mathcal{F}_1 = \bigcup_{k_1 \geq 1, k_2, k_3, k_4 \geq 0} \mathcal{T}_{k_1,k_2,k_3,k_4}.$$

A tree T in the family $\mathcal{T}_{k_1,k_2,k_3,k_4}$ is illustrated in Figure 3.2. We note that there can be additional attachments of Type-2 at each darkened vertex in Figure 3.2 that belongs to an attachment of Type-A. It is shown in [76] that the vertex a (depicted by the open square in Figure 3.2) is an optimal first move of Dominator and that every tree T in the family $\mathcal{F}_1$ satisfies $\gamma_t(T) = \gamma_{tg}(T)$.

Let F_{10} be the tree of order 10, illustrated in Figure 3.3, obtained from a star $K_{1,3}$ by subdividing two edges three times. It is shown in [76] that the two vertices x_1 and x_2 (represented by open squares) are the only optimal first moves of Dominator and $\gamma_t(F_{10}) = \gamma_{tg}(F_{10})$.

Let $\mathcal{F} = \mathcal{F}_1 \cup \{K_2, F_{10}\}$, and let $\mathcal{F}^*$ be the family of all stars on at least two vertices together with all trees that can be obtained from a tree F of order at least 3 in the family $\mathcal{F}$ by adding any number, including the possibility of zero, additional pendant edges to support vertices of F. We are now in a position to present the characterization of trees with equal total domination and game total domination numbers.

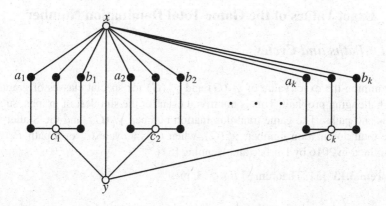

Fig. 3.4 The graph G_m.

Theorem 3.11 [76, Theorem 2] *Let T be a nontrivial tree. Then, $\gamma_t(T) = \gamma_{tg}(T)$ if and only if $T \in \mathcal{F}^*$.*

We next present a result showing that the game total domination number of a spanning tree T can be arbitrarily smaller than that of the supergraph containing T.

Theorem 3.12 *For any positive integer m there exists a graph G_m and a spanning tree T_m of G_m, such that $\gamma_{tg}(G_m) - \gamma_{tg}(T_m) \geq m$.*

Let $k = 2m + 4$ and let G_m be the graph shown in Figure 3.4. In addition, let T_m be the spanning tree of G_m obtained by deleting all of the "vertical edges" with the exception of the edge xy. For each $i \in [k]$, let $X_i = \{a_i, b_i, c_i\}$, let u_i be the common neighbor of a_i and c_i, and let v_i be the common neighbor of b_i and c_i. First we present a strategy for Dominator that will ensure that $\gamma_{tg}(T_m) \leq k + 2$. Dominator plays the vertex y as his first move. After this move no vertex that is solid-colored is playable, but all the vertices in $\{x, c_1, c_2, \ldots, c_k\}$ must be played (in any order) during the remainder of the game. Thus we see that $\gamma_{tg}(T_m) \leq k + 2$.

We now present a strategy for Staller when the D-game is played on G_m that shows $\gamma_{tg}(G_m) \geq 3k/2$. Her goal is to maximize the number of values of $i \in [k]$ such that at least two vertices from X_i are played when the game ends. During the course of the game when it is Staller's turn she searches for an $i \in [k]$ such that no vertex from X_i has already been played. If there is such an i, then Staller plays the vertex a_i. Note that this move of Staller ensures that at least one more vertex from $\{b_i, c_i\}$ is played (in order to totally dominate v_i) when the game ends. If no such i exists, then Staller plays any legal move. Since k is even, Staller can realize her goal of requiring that at least two vertices from X_i are played for at least one-half of the values in the set $[k]$. Therefore, Staller can force the game to last at least $3k/2$ moves. Hence, $\gamma_{tg}(G_m) - \gamma_{tg}(T_m) \geq \frac{3k}{2} - (k + 2) = m$.

3.5   Exact Values of the Game Total Domination Number

3.5.1   Paths and Cycles

Determining the exact value of $\gamma_{tg}(G)$ and $\gamma'_{tg}(G)$ for special classes of graphs G is a challenging problem. This is nontrivial even for the simplest of graphs, such as cycles and paths. The game total domination number, $\gamma_{tg}(G)$, and the Staller-start game total domination number, $\gamma'_{tg}(G)$, when G is a cycle C_n or a path P_n were determined in 2016 by Dorbec and Henning [51].

Theorem 3.13 [51, Theorem 3] *If $n \geq 3$, then*

$$\gamma_{tg}(C_n) = \begin{cases} \lfloor \frac{2n+1}{3} \rfloor - 1; & n \equiv 4 \,(\mathrm{mod}\ 6), \\ \lfloor \frac{2n+1}{3} \rfloor; & otherwise, \end{cases}$$

and

$$\gamma'_{tg}(C_n) = \begin{cases} \lfloor \frac{2n}{3} \rfloor - 1; & n \equiv 2 \,(\mathrm{mod}\ 6), \\ \lfloor \frac{2n}{3} \rfloor; & otherwise. \end{cases}$$

For small n, the values of $\gamma_{tg}(C_n)$ and $\gamma'_{tg}(C_n)$ are shown in Table 3.1.

Theorem 3.14 [51, Theorem 11] *If $n \geq 1$, then*

$$\gamma_{tg}(P_n) = \begin{cases} \lfloor \frac{2n}{3} \rfloor; & n \equiv 5 \,(\mathrm{mod}\ 6), \\ \lceil \frac{2n}{3} \rceil; & otherwise, \end{cases}$$

and $\gamma'_{tg}(P_n) = \lceil \frac{2n}{3} \rceil$.

For small n, the values of $\gamma_{tg}(P_n)$ and $\gamma'_{tg}(P_n)$ are shown in Table 3.2.

Table 3.1 $\gamma_{tg}(C_n)$ and $\gamma'_{tg}(C_n)$ for small cycles C_n.

n	3	4	5	6	7	8	9	10	11
$\gamma_{tg}(C_n)$	2	2	3	4	5	5	6	6	7
$\gamma'_{tg}(C_n)$	2	2	3	4	4	4	6	6	7

Table 3.2 $\gamma_{tg}(P_n)$ and $\gamma'_{tg}(P_n)$ for small paths P_n.

n	3	4	5	6	7	8	9	10
$\gamma_{tg}(P_n)$	2	3	3	4	5	6	6	7
$\gamma'_{tg}(P_n)$	2	3	4	4	5	6	6	7

3.5.2   Cyclic Bipartite Graphs

Jiang and Lu [85] considered the following class of graphs. Let $n \geq k \geq 2$ be two integers. A *cyclic bipartite graph*, $G_{n,k}$, has the vertex set $V(G_{n,k}) = \{x_i : i \in [n]\} \cup \{y_i : i \in [n]\}$ and the edge set $E(G_{n,k}) = \{x_i y_{i+j} : i \in [n], j \in [k-1]_0\}$, where the index $i + j$ is computed modulo n. Clearly, $G_{n,k}$ is a k-regular graph of order $2n$. In particular, $G_{n,2} = C_{2n}$. The following result thus generalizes Theorem 3.13 for the case of even cycles.

Theorem 3.15 [85, Theorem 2.1] *If $n \geq k \geq 2$, then*

$$\gamma_{\mathrm{tg}}(G_{n,k}) = \begin{cases} 4\ell; & n = (k+1)\ell, \\ 4\ell + 1; & n = (k+1)\ell + j, \text{ where } j \in [k-1], \\ 4\ell + 2; & n = (k+1)\ell + k, \end{cases}$$

and

$$\gamma_{\mathrm{tg}}'(G_{n,k}) = \begin{cases} 4\ell; & n = (k+1)\ell + j, \text{ where } j \in [k-1]_0, \\ 4\ell + 2; & n = (k+1)\ell + k. \end{cases}$$

3.5.3   Graphs Achieving the General Lower Bound

There are infinite families of graphs with equal total domination number and game total domination number. In Theorem 3.11 we have presented a complete characterization of the class of trees with this property. Here we describe an infinite family of general graphs. Let H be a nontrivial, connected bipartite graph with bipartition A, B having the additional property that every vertex in A is a support vertex. Let G be the graph obtained from H by adding a new vertex u and the set of edges $\{ua : a \in A\}$. It is clear that $\gamma_t(G) = 1 + |A|$. Suppose Dominator plays the vertex u as his first move when the D-game is played on G. After Dominator plays u, no vertex in B is a legal move. If $x \in A$, then x must be played in the course of the game since x is a support vertex. This shows that $\gamma_{\mathrm{tg}}(G) \leq 1 + |A| = \gamma_t(G)$, and thus by Theorem 1.3, $\gamma_t(G) = \gamma_{\mathrm{tg}}(G)$.

3.5.4   Graphs Achieving the General Upper Bound

Let G_n be the family of graphs obtained from the complete graph K on n vertices as follows. For each vertex $v \in V(K)$ take $\lceil n/2 \rceil = n'$ copies of the triangle T, denote them by $T_1^v, \ldots, T_{n'}^v$ and identify a vertex of each T_i^v with the vertex v. Thus, $n(G_n) = n + 2nn'$, $\deg_{G_n}(v) = n - 1 + 2n'$ for a vertex $v \in V(K)$, whereas $\deg_{G_n}(x) = 2$ for a vertex $x \in V(G_n) - V(K)$. For $v \in V(K)$, let $X_v = \bigcup_{i=1}^{n'} V(T_i^v)$.

Clearly, $\gamma_t(G_n) = n$, and we claim that $\gamma_{tg}(G_n) = 2n - 1$. A strategy of Staller by which we can prove $\gamma_{tg}(G_n) \geq 2n - 1$ is to select a vertex in $X_v - \{v\}$, where v is a vertex such that Dominator has not yet played any vertex in X_v. Note that until Dominator's nth move, there always exists a vertex $v \in V(K)$ such that Dominator has not yet played any vertex in X_v. Staller can thus choose a previously unchosen vertex in $X_v - \{v\}$, since every such vertex is playable before v is selected, and there are $2n' \geq n$ vertices in $X_v - \{v\}$. In other words, Staller has a legal response to the first $n - 1$ moves of Dominator, which implies that the game will last at least $2n - 1$ moves. By the upper bound in Theorem 1.3, $\gamma_{tg}(G_n) \leq 2\gamma_t(G_n) - 1 = 2n - 1$, which yields the desired equality.

3.6 Critical Graphs, Vertex Removal, and Perfect Graphs

Critical graphs with respect to the total domination game were introduced and first studied in 2017 by Henning, Klavžar, and Rall in [72]. A graph G is *total domination game critical*,[1] abbreviated γ_{tg}-critical, if $\gamma_{tg}(G) > \gamma_{tg}(G|v)$ holds for every $v \in V(G)$. The γ_{tg}-critical cycles and γ_{tg}-critical paths are characterized in [72].

Theorem 3.16 [72, Theorem 1] *If $n \geq 3$, then the cycle C_n is γ_{tg}-critical if and only if n mod $6 \in \{0, 1, 3\}$.*

Theorem 3.17 [72, Theorem 3] *If $n \geq 2$, then the path P_n is γ_{tg}-critical if and only if n mod $6 \in \{2, 4\}$.*

If G is γ_{tg}-critical and $\gamma_{tg}(G) = k$, we say that G is k-γ_{tg}-critical. The 2-γ_{tg}-critical graphs are precisely the complete graphs.

Theorem 3.18 [72, Proposition 11] *A graph is 2-γ_{tg}-critical if and only if it is a complete graph.*

If a graph contains open twins, then it is not a total domination game critical graph as first observed in [72].

Lemma 3.19 [72, Lemma 12] *If u and v are open twins in G, then $\gamma_{tg}(G) = \gamma_{tg}(G|u) = \gamma_{tg}(G|v)$.*

As an immediate consequence of Lemma 3.19, we have the following result.

Corollary 3.20 [72, Corollary 13] *Every γ_{tg}-critical graph is open twin-free.*

The following fundamental property of γ_{tg}-critical graphs is given in [72].

---

[1]In [72] the authors used the term *game total domination critical*.

Lemma 3.21 [72, Lemma 14] *If G is a γ_{tg}-critical graph and v is any vertex of G, then no neighbor of v is an optimal first move of Dominator in the D-game on $G|v$.*

A characterization of 3-γ_{tg}-critical graphs is given in [72].

Theorem 3.22 [72, Theorem 16] *Let G be a graph of order n with no isolated vertex. The graph G is 3-γ_{tg}-critical if and only if the following all hold.*

(a) *The graph G is open twin-free.*
(b) *There is no dominating vertex in G.*
(c) *For every vertex v of G of degree at most $n - 3$, there exists a vertex u of degree $n - 2$ that is not adjacent to v.*

Theorem 3.22 shows that there are infinitely many 3-γ_{tg}-critical graphs. The following result characterizes 3-γ_{tg}-critical graphs that can be obtained from the join of two graphs.

Theorem 3.23 [72, Theorem 17] *If G_1 and G_2 are vertex disjoint graphs, then $G_1 + G_2$ is 3-γ_{tg}-critical if and only if for each $i \in [2]$ one of the following holds.*

(a) *The graph G_i is 3-γ_{tg}-critical.*
(b) *The graph G_i is $K_1 \sqcup K_k$ for some $k \geq 2$.*

Infinite families of circular and Möbius ladders that are total domination game critical are studied by Henning and Klavžar [69]. The *circular ladder graph* CL_n of order $2n$ is the Cartesian product of a cycle C_n on $n \geq 3$ vertices and a path P_2 on two vertices; that is, $CL_n = C_n \,\square\, K_2$. We note that CL_n is bipartite if and only if n is even. The circular ladders CL_4 and CL_8 are illustrated in Figs. 3.5(a) and 3.5(b), respectively.

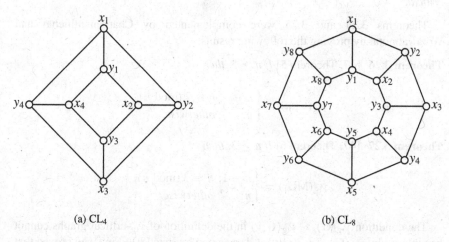

(a) CL_4 (b) CL_8

Fig. 3.5 The circular ladders CL_4 and CL_8.

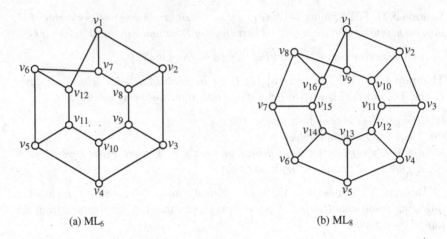

(a) ML₆   (b) ML₈

Fig. 3.6 The Möbius ladders ML₆ and ML₈.

Theorem 3.24 [69, Theorem 1] *If $k \geq 1$, then the circular ladder CL_{4k} is $4k$-γ_{tg}-critical.*

For $n \geq 2$, the Möbius ladder ML_n is a cubic graph of order $2n$, formed from a $2n$-cycle by adding n edges (called "rungs") joining opposite pairs of vertices in the cycle. The Möbius ladder ML_2 is the complete graph K_4. The Möbius ladders ML_6 and ML_8 are illustrated in Figure 3.6(a) and 3.6(b), respectively. The Möbius ladder ML_n is bipartite if and only if n is odd.

Theorem 3.25 [69, Theorem 2] *If $k \geq 1$, then the Möbius ladder ML_{2k} is $2k$-γ_{tg}-critical.*

Theorems 3.24 and 3.25 were complemented by Charoensitthichai and Worawannotai by proving the following results.

Theorem 3.26 [47, Theorem 5] *If $n \geq 3$, then*

$$\gamma_{tg}(CL_n) = \begin{cases} n-1; & n \equiv 2 \pmod{n}, \\ n; & otherwise. \end{cases}$$

Theorem 3.27 [47, Theorem 6] *If $n \geq 3$, then*

$$\gamma_{tg}(ML_n) = \begin{cases} n-1; & n \equiv 3 \pmod{n}, \\ n; & otherwise. \end{cases}$$

The condition $\gamma_{tg}(G) > \gamma_{tg}(G|v)$ in the definition of γ_{tg}-critical graphs cannot be replaced by $\gamma_{tg}(G) = \gamma_{tg}(G|v) + 1$, as noted by Iršič [80]. First, she proved that $\gamma_{tg}(G) - \gamma_{tg}(G|v) \leq 2$ holds for all graphs G with no isolated vertices [80, Lemma 2.1]. Second, she constructed an infinite family of graphs G for which $\gamma_{tg}(G) =$

$\gamma_{tg}(G|v) + 2$ as follows. Let $n \geq 8$ and $m \geq 4$ be positive integers such that $n \equiv 2 \pmod 6$, let C be the n-cycle with vertices $u_1, \ldots, u_n$ and edges in the natural order, and let K be the m-clique in which we distinguish two vertices v and w. The graph $G_{n,m}$ is obtained from the disjoint union of C and K by adding edges $u_1 v$, $u_1 w$, $u_5 v$, and $u_5 w$.

Theorem 3.28 [80, Theorem 2.2] *If $n \geq 8$ and $m \geq 4$ are positive integers such that $n \equiv 2 \pmod 6$, then $\gamma_{tg}(G_{n,m}) = \frac{2n-1}{3} + 2$ and $\gamma_{tg}(G_{n,m}|v) = \frac{2n-1}{3}$.*

There are many open problems on γ_{tg}-critical graphs; here is a selection of them.

Problem 3.29 Characterize the k-γ_{tg}-critical graphs for $k \geq 4$.

Problem 3.30 Characterize the γ_{tg}-critical trees.

Problem 3.31 Find infinite classes of γ_{tg}-critical graphs different from circular ladders and Möbius ladders.

Another graph operation that has been studied with respect to the game total domination number is vertex removal. Like in the domination game, the removal of a vertex can result in an increase of the game total domination number by an arbitrary amount. For instance, if D_n is the graph obtained from n copies of the triangle by identifying a vertex of each copy to the single vertex v, then $\gamma_{tg}(D_n) = 2$, while $\gamma_{tg}(D_n - v) = 2n$. An analogous observation holds for the S-game, since $\gamma'_{tg}(D_n) = 2$, while $\gamma'_{tg}(D_n - v) = 2n$.

On the other hand, if G is a graph with no isolated vertices and G has a vertex v such that $G - v$ has no isolated vertices, then the difference $\gamma_{tg}(G) - \gamma_{tg}(G - v)$ is bounded from above. This fact was first observed by Iršič [80, Proposition 4.1], and her result was later improved by Charoensitthichai and Worawannotai [48].

Theorem 3.32 [48, Theorem 3.1] *If G is a graph with no isolated vertices and v is a vertex in G such that $G - v$ has no isolated vertices, then $\gamma_{tg}(G) - \gamma_{tg}(G - v) \leq 2$.*

A similar result holds for the S-game.

Theorem 3.33 [48, Theorem 3.2] *If G is a graph with no isolated vertices and v is a vertex in G such that $G - v$ has no isolated vertices, then $\gamma'_{tg}(G) - \gamma'_{tg}(G - v) \leq 2$.*

Theorems 3.32 and 3.33 are sharp in a strong sense, which is presented in the following two results by Charoensitthichai and Worawannotai [48].

Theorem 3.34 [48, Theorem 4.1] *If $a \geq 2$ and $b \geq 2$ are positive integers with $a - b \leq 2$, then there exists a graph G with a vertex v such that $(\gamma_{tg}(G), \gamma_{tg}(G - v)) = (a, b)$ except for $(a, b) = (4, 2)$.*

Theorem 3.35 [48, Theorem 4.2] *If $a \geq 2$ and $b \geq 2$ are positive integers with $a - b \leq 2$, then there exists a graph G with a vertex v such that $(\gamma'_{tg}(G), \gamma'_{tg}(G - v)) = (a, b)$ except for $(a, b) = (3, \ell)$ for all $\ell \geq 6$.*

It is interesting that the effect of edge removal for the total domination game has not yet been established, while Theorem 2.47 gives the complete answer for the related problem in the domination game. It is possible that this is much harder in the total version of the game, and we propose it as the following problem.

Problem 3.36 Find the smallest possible positive integers s and t (if they exist) such that for every graph G and any edge $e \in E(G)$ such that $G - e$ has no isolated vertices,

$$\gamma_{tg}(G) - \gamma_{tg}(G - e) \leq s \text{ and } \gamma_{tg}(G - e) - \gamma_{tg}(G) \leq t.$$

In Section 2.10 we have considered γ_g- and γ_g'-perfect graphs. Analogous concepts can be defined also for the total version of the game. More precisely, call G to be γ_{tg}-*minimal* if $\gamma_t(G) = \gamma_{tg}(G)$ holds. Then G is γ_{tg}-*perfect* if all of its isolate-free induced subgraphs are γ_{tg}-minimal graphs. Similarly, G is a γ_{tg}'-*minimal graph* if $\gamma_t(G) = \gamma_{tg}'(G)$ holds, and is γ_{tg}'-*perfect* if all of its isolate-free induced subgraphs are γ_{tg}'-minimal graphs. Proofs of the following results by Bujtás, Iršič, and Klavžar are significantly simpler than the proof of the corresponding Theorem 2.61.

Proposition 3.37 [37, Proposition 5.2] *An isolate-free graph is γ_{tg}-perfect if and only if it is $(P_4, \overline{2P_3})$-free.*

Proposition 3.38 [37, Proposition 5.3] *An isolate-free graph is γ_{tg}'-perfect if and only if it is a cograph.*

3.7 Computational Complexity

The algorithmic complexity of determining the game total domination number of a given graph was studied by Brešar and Henning [20]. Formally, they consider the following game total domination problems.

D-GAME TOTAL DOMINATION PROBLEM

Input: A graph G, and an integer ℓ.
Question: Is $\gamma_{tg}(G) \leq \ell$?

S-GAME TOTAL DOMINATION PROBLEM

Input: A graph G, and an integer ℓ.
Question: Is $\gamma_{tg}'(G) \leq \ell$?

By using a similar setting as in Section 2.12, a reduction to the game total domination problems from the POS-CNF problem is presented. For this purpose, a gadget graph T with specified vertex a shown in Figure 3.7 is defined.

Fig. 3.7 The gadget, T, used in the construction, representing a variable.

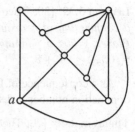

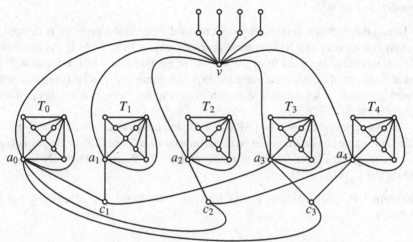

Fig. 3.8 The graph, G_F, associated with the formula $\mathcal{F} = (X_1 \vee X_3) \wedge (X_2 \vee X_4) \wedge (X_3 \vee X_4)$.

Given a formula $\mathcal{F}$ using k variables and n disjunctive clauses, the authors in [20] build a graph $G_{\mathcal{F}}$, having $8(k+1) + n + 9$ vertices, as follows. For each variable X_i (where $i \in [k]$) we add to the graph a copy T_i of the gadget graph T shown in Figure 3.7, and add an additional copy T_0 of T. Let a_i be the vertex in T_i corresponding to the vertex a in the gadget graph T where $i \in [k]_0$. For each clause C_j (where $j \in [n]$) we add a vertex c_j to the graph, and make c_j adjacent to a vertex a_i, whenever the variable X_i appears in the clause C_j. We join the vertex a_0 in the gadget T_0 to all vertices c_j, for $j \in [n]$. Finally, we add a graph obtained from a star $K_{1,4}$ centered at a vertex v by subdividing every edge of the star exactly once, and join v to all vertices a_i, where $i \in [k]_0$.

To illustrate the construction of the graph $G_{\mathcal{F}}$, consider the formula $\mathcal{F} = C_1 \wedge C_2 \wedge C_3$, where $C_1 = X_1 \vee X_3$, $C_2 = X_2 \vee X_4$, and $C_3 = X_3 \vee X_4$. In this case, there are $k = 4$ variables and $n = 3$ disjunctive clauses. The associated graph $G_{\mathcal{F}}$ is illustrated in Figure 3.8, where in each gadget, T_i, where $i \in [4]_0$, we only label the vertex a_i (corresponding to the vertex a in the gadget T).

The following result establishes a connection between the game total domination problems and the POS-CNF problem.

Lemma 3.39 [20, Theorem 3] *Let $\mathcal{F}$ be a formula with k variables, which is a conjunction of n disjunctive clauses, and let $G_{\mathcal{F}}$ be the corresponding graph. Player 1 has a winning strategy for $\mathcal{F}$ in the POS-CNF game if and only if $\gamma_{\mathrm{tg}}(G_{\mathcal{F}}) \leq 3k + 8$.*

As a consequence of Lemma 3.39, the algorithmic complexity of the game total domination number is determined.

Theorem 3.40 [20, Theorem 2] D-GAME TOTAL DOMINATION PROBLEM *is log-complete in PSPACE.*

Hence the decision version of the game total domination problem is computationally harder than any NP-complete problem, unless NP=PSPACE. As remarked in [20], the reduction used from the POS-CNF problem, in which a formula $\mathcal{F}$ is given, to the total domination game of a special graph $G_{\mathcal{F}}$ can be computed with a working space of logarithmic size with respect to the entry, making this problem log-complete in PSPACE.

An analogous result, that is, PSPACE-completeness, is true also for the S-game. Indeed the same construction can be used, translating a formula $\mathcal{F}$ to the graph $G_{\mathcal{F}}$, and showing that Player 1 has a winning strategy for $\mathcal{F}$ in the POS-CNF game if and only if $\gamma'_{\mathrm{tg}}(G_{\mathcal{F}}) \leq 3k + 9$.

Theorem 3.41 [20] S-GAME TOTAL DOMINATION PROBLEM *is log-complete in PSPACE.*

Chapter 4
Games for Staller

The following question was posed by Brešar, Gologranc, Milanič, Rall, and Rizzi [19]. What if the domination game is played by only one of the players using his or her goal? Naturally, if Dominator is the only one to play a domination game in a graph G, the resulting number of moves will be the domination number $\gamma(G)$. A more interesting version of this question is: What happens when only Staller is playing the game? The corresponding graph invariant, called the Grundy domination number, thus presents the worst outcome of a greedy algorithm for constructing a dominating set if one is processing the addition of vertices one at a time.[1]

The Grundy domination number is presented in Section 4.1. In Section 4.2, we consider the total version of this concept, which corresponds to the total domination game played only by Staller. We follow with a common generalization of both types of Grundy domination numbers to hypergraphs, which is applied in determining the computational complexity of all three problems. Section 4.4 is concerned with graphs whose Grundy (total) domination number equals their domination number, which brings along surprising connections with some well-known discrete structures. In addition, connections with linear algebra are established. Notably, in Section 4.5 a variation of the condition for adding a vertex while building a dominating set results in an invariant, which is dual to the well-known zero forcing number. The resulting invariant is called the Z-Grundy domination number. Then, we present yet another natural variation, giving the fourth Grundy-type domination number, and compare all four invariants among themselves. In Section 4.8, we consider complexity issues of these invariants, and at the end present what has been done with respect to various classes of graphs; either exact values or algorithms for their Grundy domination numbers.

---

[1]Note that the Grundy domination number is related to the domination number in a similar way as the Grundy (chromatic) number is related to the chromatic number.

B. Brešar et al., *Domination Games Played on Graphs*, SpringerBriefs in Mathematics, https://doi.org/10.1007/978-3-030-69087-8_4

4.1   The Grundy Domination Number

Let $S = (v_1, \ldots, v_k)$ be a sequence of distinct vertices of a graph G, and denote the corresponding set $\{v_1, \ldots, v_k\}$ of vertices from the sequence S by $\widehat{S}$. A sequence $S = (v_1, \ldots, v_k)$, where $v_i \in V(G)$, is a *closed neighborhood sequence* if

$$N[v_{i+1}] - \bigcup_{j=1}^{i} N[v_j] \neq \emptyset \qquad (4.1)$$

for each $i \in [k-1]$. If, in addition, $\widehat{S}$ is a dominating set of G, then S is a *dominating sequence* in G. We will also say that v_{i+1} *footprints* the vertices from $N[v_{i+1}] - \bigcup_{j=1}^{i} N[v_j]$, and that v_{i+1} is the *footprinter* of any $u \in N[v_{i+1}] - \bigcup_{j=1}^{i} N[v_j]$. For a dominating sequence S, any vertex in $V(G)$ has a unique footprinter in $\widehat{S}$, which makes the function $f_S : V(G) \to \widehat{S}$ that maps each vertex to its footprinter well defined. Clearly, a shortest possible dominating sequence has length $\gamma(G)$ (it is in the game played only by Dominator). A longest possible dominating sequence in G is called a *Grundy dominating sequence*, and its length is the *Grundy domination number* of G, denoted $\gamma_{gr}(G)$.

It is obvious that for every graph G,

$$\gamma(G) \leq \gamma_g(G) \leq \gamma_{gr}(G), \qquad (4.2)$$

and, in addition, the Grundy domination number in any graph is at least as large as most of the other domination-type invariants, such as the upper domination number and the upper irredundance number.

We start with the following general upper bound on the Grundy domination number.

Proposition 4.1 [19, Proposition 2.1] *If G is a graph, then $\gamma_{gr}(G) \leq n(G) - \delta(G)$.*

Proof. Let $(s_1, \ldots, s_k)$ be a Grundy dominating sequence of G, and let u be a vertex footprinted by s_k. By definition of the footprinting function, $N[u] \cap \{s_1, \ldots, s_{k-1}\} = \emptyset$, which gives $|\{s_1, \ldots, s_{k-1}\}| = k - 1 \leq n(G) - (\deg(u) + 1)$. Thus, $\gamma_{gr}(G) = k \leq n(G) - \delta(G)$. $\square$

The bound in Proposition 4.1 is attained by a number of graphs families (examples are complete graphs, complete bipartite graphs, and caterpillars).

For the next upper bound, established by Brešar, Bujtás, Gologranc, Klavžar, Košmrlj, Patkós, Tuza, and Vizer, we recall the following invariant on a graph G. A set of complete subgraphs $Q_1, \ldots, Q_r$ in G such that for every edge $e \in E(G)$ there exists an $i \in [r]$ such that $e \in E(Q_i)$ is an *edge clique cover* of G. The minimum cardinality of an edge-clique cover in G is the *edge clique cover number*, $\theta_e(G)$, of G.

Proposition 4.2 [12, Proposition 2] *If G is a graph without isolated vertices, then* $\gamma_{gr}(G) \leq \theta_e(G)$.

Proof. Let Q be a minimum edge clique cover of G and let $(s_1, \ldots, s_k)$ be a Grundy dominating sequence of G. We claim that each time a vertex s_i is selected, all vertices of some clique in Q become dominated. Indeed, if s_i footprints its neighbor u, then the edge $s_i u$ lies in a clique $Q_j \in Q$. Before s_i was selected, u had not been dominated by $\widehat{S_{i-1}}$, and after s_i is selected, all vertices of Q_j are dominated. On the other hand, if s_i footprints only itself, then for any $u \in N_G(s_i)$ the edge $s_i u$ lies in a clique of Q all of whose vertices become dominated only after s_i is selected. Consequently, Q contains at least k cliques, that is, $\theta_e(G) = |Q| \geq k = \gamma_{gr}(G)$. $\square$

The bound in Proposition 4.2 is sharp, as demonstrated by complete graphs K_n and caterpillars. In fact, as observed in [12], in the class of trees T, the bound is sharp precisely when T is a caterpillar.

The effect of edge or vertex deletion on the Grundy domination number of a graph was considered by Brešar, Gologranc, and Kos [18].

Theorem 4.3 [18, Theorem 1] *If G is a graph and $e \in E(G)$, then*

$$\gamma_{gr}(G) - 1 \leq \gamma_{gr}(G - e) \leq \gamma_{gr}(G) + 1.$$

Proof. Let $e = uv$ be the edge deleted from a graph G. To prove $\gamma_{gr}(G) - 1 \leq \gamma_{gr}(G - e)$, consider a Grundy dominating sequence S in G. If neither of u and v is in $\widehat{S}$, then S is also a closed neighborhood sequence in $G - e$, which implies $\gamma_{gr}(G - e) \geq |\widehat{S}| = \gamma_{gr}(G)$. Suppose $u \in \widehat{S}$ and let u be the first among u and v that appears in S (this includes the case when $v \notin \widehat{S}$). Then the sequence S' obtained from S by skipping u is a closed neighborhood sequence in $G - e$, which implies $\gamma_{gr}(G - e) \geq |\widehat{S'}| = \gamma_{gr}(G) - 1$.

For the other inequality let $S = (x_1, \ldots, x_k)$ be a Grundy dominating sequence in $G - e$ and let f_S be the corresponding footprinting function on $V(G - e)$. If neither u nor v is in S, then S is a closed neighborhood sequence in G, and thus $\gamma_{gr}(G) \geq \gamma_{gr}(G - e)$. Suppose $u \in \widehat{S}$ is the first among u and v to appear in S (including the case when $v \notin \widehat{S}$), and let u be the rth vertex in S, that is, $u = x_r$. Note that for all $i \in [r]$,

$$N_G[x_i] - \bigcup_{j=1}^{i-1} N_G[x_j] \neq \emptyset. \tag{4.3}$$

Suppose that $f_S(v) = x_s$ with $s > r$, and $f_S^{-1}(x_s) = \{v\}$. (If this is not the case, then S is a closed neighborhood sequence also in G.) In that case, the condition (4.3) does not hold for $i = s$. However, removing x_s from S, the condition (4.3) holds for all vertices x_i, where $i \in \{r+1, \ldots, k\} - \{s\}$. In other words, the resulting sequence of length $k - 1$ is a closed neighborhood sequence in G, hence $\gamma_{gr}(G) \geq \gamma_{gr}(G-e) - 1$. $\square$

It was also proven in [18] that there exists an infinite family of graphs G with edges e_1, e_2, e_3 such that $\{\gamma_{gr}(G - e_1), \gamma_{gr}(G - e_2), \gamma_{gr}(G - e_3)\} = \{\gamma_{gr}(G) - 1, \gamma_{gr}(G), \gamma_{gr}(G) + 1\}$.

If S is a closed neighborhood sequence in a graph H that is an induced subgraph of a graph G, then S is also a closed neighborhood sequence in G. This immediately implies $\gamma_{gr}(H) \leq \gamma_{gr}(G)$ and proves the upper bound in the following theorem.

Theorem 4.4 [18, Theorem 3] *If G is a graph and $v \in V(G)$, then*

$$\gamma_{gr}(G) - 2 \leq \gamma_{gr}(G - v) \leq \gamma_{gr}(G).$$

Similarly as for edge-deletion, it was proven in [18] that there exists an infinite family of graphs G with vertices v_1, v_2, v_3 such that $\{\gamma_{gr}(G - v_1), \gamma_{gr}(G - v_2), \gamma_{gr}(G - v_3)\} = \{\gamma_{gr}(G) - 2, \gamma_{gr}(G) - 1, \gamma_{gr}(G)\}$. In addition, if v is a simplicial vertex, then $\gamma_{gr}(G - v) \geq \gamma_{gr}(G) - 1$, and if v is a twin vertex, then $\gamma_{gr}(G - v) = \gamma_{gr}(G)$; see [18, Proposition 4].

Motivated by results from [21], Brešar and Brezovnik [9] considered Grundy domination numbers in regular graphs.

Theorem 4.5 [9, Theorem 2.1] *If $k \geq 3$ and G is a connected k-regular graph of order n different from K_{k+1} and $\overline{C_4 \sqcup C_4}$, then $\gamma_{gr}(G) \geq \frac{n + \lceil \frac{k}{2} \rceil - 2}{k - 1}$.*

In cubic graphs G, the above result implies the bound $\gamma_{gr}(G) \geq n(G)/2$, and cubic graphs with $\gamma_{gr}(G) = n(G)/2$ were characterized in [9, Corollary 4.6].

4.2 The Grundy Total Domination Number

Brešar, Henning, and Rall [21] introduced the total version of Grundy domination, which arises when Staller is the only player in the total domination game. We assume that graphs in this section have no isolated vertices even if this is not always stated explicitly.

Let G be a graph (with no isolated vertices). A sequence $S = (v_1, \ldots, v_k)$ of (distinct) vertices of G is an *open neighborhood sequence* if

$$N(v_{i+1}) - \bigcup_{j=1}^{i} N(v_j) \neq \emptyset \tag{4.4}$$

for each $i \in [k - 1]$. If, in addition, $\widehat{S}$ is a total dominating set of G, then S is a *total dominating sequence* in G. We will also say that v_{i+1} *t-footprints* the vertices from $N(v_{i+1}) - \bigcup_{j=1}^{i} N(v_j)$, and that v_{i+1} is the *t-footprinter* of any $u \in N(v_{i+1}) - \bigcup_{j=1}^{i} N(v_j)$. For a total dominating sequence S, any vertex in $V(G)$ has a unique t-footprinter in $\widehat{S}$, which makes the function $f_S : V(G) \to \widehat{S}$ that maps each vertex to its t-footprinter well defined. A shortest possible total dominating sequence has length $\gamma_t(G)$. A longest possible total dominating sequence in G (which coincides with the number of moves in the total domination game played only by Staller) is called a *Grundy total dominating sequence*, and its length is the *Grundy total domination number* of G, denoted by $\gamma_{gr}^t(G)$. The following inequalities, analogous to those in Equation (4.2), hold:

$$\gamma_t(G) \leq \gamma_{tg}(G) \leq \gamma_{gr}^t(G). \tag{4.5}$$

Note that $\gamma_{gr}(G) = \gamma(G) = 1$ if G is a complete graph. Moreover, for any integer k, there exists a graph G with $\gamma_{gr}(G) = k$. On the other hand, it is clear that $\gamma_{gr}^t(G)$ is always at least 2. As it turns out, $\gamma_{gr}^t(G) = 2$ if and only if G is a complete multipartite graph [21, Theorem 4.4]. Somewhat surprisingly there are no graphs with Grundy total domination number equal to 3, as shown by Brešar, Kos, Nasini, and Torres [26].

Theorem 4.6 [26, Proposition 3.9 and Theorem 3.10] *There exists no graph G such that $\gamma_{gr}^t(G) = 3$. For every positive integer k not equal to 1 or 3, there exists a graph G with $\gamma_{gr}^t(G) = k$.*

The first statement of Theorem 4.6 can be proven as follows. Suppose that G is a graph with a Grundy total dominating sequence (s_1, s_2, s_3). Let t_i be t-footprinted by s_i for $i \in [3]$. If s_1 and s_2 are not adjacent, then (s_1, s_2, t_2, t_1) is an open neighborhood sequence, which yields $\gamma_{gr}^t(G) \geq 4$, a contradiction. On the other hand, if s_1 and s_2 are adjacent, then note that (t_3, s_1, s_2, s_3) is an open neighborhood sequence. Indeed, t_3 t-footprints s_3, s_1 t-footprints s_2, s_2 t-footprints s_1, and s_3 t-footprints t_3. This is again a contradiction. For the proof of the second statement of Theorem 4.6, see [26].

Relating both invariants, we first observe that the Grundy domination number can be arbitrarily larger than the Grundy total domination number. For instance, $\gamma_{gr}^t(K_{r,s}) = 2$, while $\gamma_{gr}(K_{r,s}) = \max\{r, s\}$. (This is in contrast to the fact that $\gamma_g(G) \leq 2\gamma_{tg}(G) - 1$ holds for every isolate-free graph G; see Chapter 3.) On the other hand, for every graph G with no isolated vertices, we have $\gamma_{gr}^t(G) \leq 2\gamma_{gr}(G)$; see [21, Theorem 7.1].

We continue with an upper bound whose proof is similar to the proof of Proposition 4.1.

Proposition 4.7 [21, Corollary 4.3] *If G is a graph with no isolated vertices, then $\gamma_{gr}^t(G) \leq n(G) - \delta(G) + 1$.*

Unlike with the Grundy domination number, which in non-trivial graphs cannot be equal to the order, the bound in Proposition 4.7 shows that for the Grundy total domination number this is possible only when a graph has a vertex of degree 1. The family of graphs G with $\gamma_{gr}^t(G) = n(G)$ is in fact rather rich.

Theorem 4.8 *[21, Theorem 4.2] If G is a graph, then $\gamma_{gr}^t(G) = n(G)$ if and only if there exists an integer k such that $n(G) = 2k$, and the vertices of G can be labeled $x_1, \ldots, x_k, y_1, \ldots, y_k$ in such a way that*

- x_i *is adjacent to* y_i *for each i,*
- $\{x_1, \ldots, x_k\}$ *is an independent set, and*
- y_j *is adjacent to* x_i *implies* $i \geq j$.

By Theorem 4.8, any graph G with $\gamma_{gr}^t(G) = n(G)$ has a perfect matching. In fact, for trees this is also a sufficient condition; see [21, Theorem 5.1].

Grundy total domination was also studied by Brešar, Bujtás, Gologranc, Klavžar, Košmrlj, Marc, Patkós, Tuza, and Vizer [10]. They presented the following general upper bound for the Grundy total domination number of a graph G in terms of the Grundy total domination numbers for some collection of subgraphs of G.

Lemma 4.9 *[10, Lemma 2.3] Let $E_1, \ldots, E_k$ be non-empty subsets of the edge set $E(G)$ of a graph G such that $E_1 \cup \cdots \cup E_k = E(G)$. If, for each $i \in [k]$, G_i is the subgraph of G induced by E_i, then*

$$\gamma_{gr}^t(G) \leq \gamma_{gr}^t(G_1) + \cdots + \gamma_{gr}^t(G_k).$$

We get interesting consequences of Lemma 4.9 when we restrict to subgraphs induced by the edges of complete bipartite graphs. Let $bc(G)$ be the smallest size of a covering of the edges of G with complete bipartite graphs. (Note that $\gamma_{gr}^t(K_{r,s}) = 2$.)

Corollary 4.10 *[10, Corollary 2.4] If G is a graph, then $\gamma_{gr}^t(G) \leq 2bc(G)$.*

A special case of Corollary 4.10 is obtained when we further restrict the complete bipartite subgraphs to those with a dominating vertex, that is, to stars. It is easy to see that the smallest size of a covering of the edges of G with stars equals the vertex cover number $\beta(G)$. This gives an upper bound for the Grundy total domination number, weaker than the one in Corollary 4.10, notably $\gamma_{gr}^t(G) \leq 2\beta(G)$. However, if T is a forest with non-trivial components, then $\gamma_{gr}^t(T) = 2\beta(T)$, as shown in [26, Theorem 5.1].

A matching M is called a *semistrong* matching if every edge in M has a vertex of degree 1 in the subgraph induced by the edges of M. The number of edges in a maximum semistrong matching of G is the *semistrong matching number*, $\alpha_{ss}'(G)$, of G.

Proposition 4.11 *[21, Corollary 4.3] If G is a graph with no isolated vertices, then $\gamma_{gr}^t(G) \geq 2\alpha_{ss}'(G)$.*

Note that the bound in Proposition 4.11 is sharp, since in the corona $H \odot K_1$ of an arbitrary graph H, we have $\gamma_{gr}^t(H \odot K_1) = 2\alpha_{ss}'(H \odot K_1) = 2n(H)$.

The effect of edge or vertex deletion on the Grundy total domination number of a graph was also considered. First of all note that if v is an open twin of a graph G, then $\gamma_{gr}^t(G - v) = \gamma_{gr}^t(G)$.

Proposition 4.12 [26, Lemma 3.5] *If G is an isolate-free graph and $v \in V(G)$ is not a support vertex, then*

$$\gamma_{\mathrm{gr}}^t(G) - 2 \le \gamma_{\mathrm{gr}}^t(G - v) \le \gamma_{\mathrm{gr}}^t(G),$$

and the bounds are sharp.

Proof. For the proof of the upper bound note that if S is a Grundy total dominating sequence in $G - v$, then S is an open neighborhood sequence in G, which implies $\gamma_{\mathrm{gr}}^t(G) \ge |\widehat{S}| = \gamma_{\mathrm{gr}}^t(G - v)$. Now, let S' be a Grundy total dominating sequence in G, and let $u \in \widehat{S'}$ be a vertex that t-footprints v. Note that the sequence obtained from S' by removing u, and also removing v if $v \in \widehat{S'}$, is an open neighborhood sequence in $G - v$. Thus, $\gamma_{\mathrm{gr}}^t(G - v) \ge |\widehat{S'}| - 2 = \gamma_{\mathrm{gr}}^t(G) - 2$. The lower bound is sharp, as demonstrated by the corona $P_n \odot K_1$, where v is a leaf adjacent to a vertex of degree 2. Indeed, $\gamma_{\mathrm{gr}}^t(P_n \odot K_1) = 2n = \gamma_{\mathrm{gr}}^t(P_n \odot K_1 - v) + 2$. The upper bound is obtained in graphs with open twins as remarked before the statement of this result. □

In his doctoral dissertation, Kos [90] proved upper and lower bounds on the Grundy total domination number when a single edge is deleted.

Theorem 4.13 [90, Theorem 3.24] *If G is a graph and $e \in E(G)$, then*

$$\gamma_{\mathrm{gr}}^t(G) - 2 \le \gamma_{\mathrm{gr}}^t(G - e) \le \gamma_{\mathrm{gr}}^t(G) + 2,$$

and the bounds are sharp.

The lower bound in Theorem 4.13 is attained by the even path P_{4k+2}, where $k \ge 1$, by removing the edge e joining the two central vertices. The upper bound is attained by the even cycle C_{2k}.

There is no general lower bound for the Grundy total domination number of a graph in terms of its order when graphs with open twins are considered. For instance, $\gamma_{\mathrm{gr}}^t(K_{m,n}) = 2$. Restricting to trees having no open twins (which is equivalent to saying they have no strong support vertices), the following lower bound in terms of the order was proved.

Theorem 4.14 [21, Theorem 5.4] *If T is a tree with no strong support vertex, then* $\gamma_{\mathrm{gr}}^t(T) \ge \frac{2}{3}(n(T) + 1)$.

In addition, Brešar, Henning, and Rall in [21] characterized the class of trees that attain the lower bound of Theorem 4.14. In the same paper, a lower bound for the Grundy total domination number was proved for regular graphs.

Theorem 4.15 [21, Theorem 6.3] *If $k \ge 3$ and G is a connected k-regular graph of order n different from $K_{k,k}$, then*

$$\gamma_{\mathrm{gr}}^t(G) \ge \begin{cases} \dfrac{n + \lceil \frac{k}{2} \rceil - 2}{k - 1}; & G \text{ not bipartite,} \\[2mm] \dfrac{n + 2\lceil \frac{k}{2} \rceil - 4}{k - 1}; & G \text{ bipartite.} \end{cases}$$

4.3   Covering Sequences in Hypergraphs

Given a (finite) hypergraph $H = (V, F)$, an *edge cover* of H is a set of edges F' in F such that for every vertex $x \in V$ there exists an edge in F' containing x. The *edge covering number* of H, $\rho(H)$, is the cardinality of a minimum edge cover.

If, in the process of building an edge cover of a hypergraph H, edges are added on-line, one by one, the resulting sequence $(f_1, \ldots, f_k)$ is a *covering sequence* in H if every edge f_i contains a vertex x such that $x \notin \cup_{j=1}^{i-1} f_j$. While the length of a shortest covering sequence in H is $\rho(H)$, we call the maximum length of a covering sequence the *Grundy covering number* of H, and denote it by $\rho_{\mathrm{gr}}(H)$.

The concept of Grundy covering sequences in hypergraphs was introduced by Brešar, Gologranc, Milanič, Rall, and Rizzi [19], and presents a common generalization of Grundy dominating and Grundy total dominating sequences. This generalization comes from two natural hypergraphs that are derived from an arbitrary graph G, namely, the closed and the open neighborhood hypergraph of G. Indeed there is a bijection between closed (open) neighborhood sequences in G and covering sequences in the closed (open) neighborhood hypergraph of G. In particular, if H is the closed neighborhood hypergraph of a graph G, then $\rho(H) = \gamma(G)$ and $\rho_{\mathrm{gr}}(H) = \gamma_{\mathrm{gr}}(G)$. Similar statements hold for the open neighborhood hypergraph of G.

Let $H = (V, F)$ be a hypergraph. A covering sequence $(f_1, \ldots, f_k)$, where $f_i \in F$, is *commutative* if the sequence formed by any permutation of its edges is also a covering sequence. We say that an edge cover Q of H is a *minimal edge cover* if no proper subset of Q is an edge cover.

Proposition 4.16 [19, Lemma 3.1] *A covering sequence in a hypergraph is commutative if and only if the family of the edges in this sequence is a minimal edge cover.*

By Proposition 4.16, we infer that a closed (open) neighborhood sequence S in a graph is commutative if and only if $\widehat{S}$ is a minimal (total) dominating set. The following result about covering sequences in hypergraphs also has immediate consequences for dominating and total dominating sequences.

Theorem 4.17 [19, Theorem 3.2] *Let H be a hypergraph. There exists a covering sequence in H having length ℓ, for every $\ell \in \{\rho(H), \ldots, \rho_{\mathrm{gr}}(H)\}$.*

Covering sequences in hypergraphs were used in [19, 21] to establish the computational complexity of decision versions of the Grundy domination number and the Grundy total domination number, respectively. The results were based on the following decision version of the Grundy covering problem in hypergraphs.

GRUNDY COVERING NUMBER IN HYPERGRAPHS

Input: A hypergraph $H = (X, F)$, and an integer k.
Question: Is $\rho_{\mathrm{gr}}(H) \geq k$?

GRUNDY COVERING NUMBER IN HYPERGRAPHS is clearly in NP, while the NP-completeness was proven by a reduction from FEEDBACK ARC SET problem.

Theorem 4.18 [19, Theorem 4.2] GRUNDY COVERING NUMBER IN HYPER-GRAPHS *is NP-complete.*

For the proof of the complexity result in [21], the following natural concept was also used, which is dual to the Grundy covering number. Let $H = (V, F)$ be a hypergraph. A sequence $S = (v_1, \ldots, v_k)$ of vertices in H is a *transversal sequence* if for each $i \in [k]$ there exists an edge $E_i \in F$ such that $v_i \in E_i$, and $v_j \notin E_i$ for all $j \in [i - 1]$. The *Grundy transversal number* of H, $\tau_{\mathrm{gr}}(H)$, is the maximum length of a transversal sequence in H. Brešar, Henning, and Rall proved that the Grundy covering number and the Grundy transversal number of a hypergraph are equal.

Proposition 4.19 [21, Proposition 8.3] *If H is a hypergraph, then $\tau_{\mathrm{gr}}(H) = \rho_{\mathrm{gr}}(H)$.*

Jayaram, Arumugam, and Thulasiraman [84] introduced the concept of *dominator sequences in bipartite graphs*. In a bipartite graph G with partite sets X and Y, where $|X| \leq |Y|$, they considered a sequence $(v_1, \ldots, v_k)$ of vertices in X with the property that $N(v_i) - \cup_{j=1}^{i-1} N(v_j) \neq \emptyset$ for all $i \in [k]$. Note that this concept coincides with a covering sequence of the hypergraph $(Y, \{N(x)\}_{x \in X})$. Beside a number of basic results on dominator sequences, they also present an application in optical networks; see [84].

4.4 Uniform Dominating Sequence Graphs

In this section, we consider two special classes of graphs for which the invariants in the inequality chains of (4.2) and (4.5), respectively, are equal. A graph G is a *k-uniform dominating sequence graph*, abbreviated *k-UDS graph*, if $\gamma(G) = \gamma_{\mathrm{gr}}(G) = k$. Similarly, G is a *k-uniform total dominating sequence graph*, abbreviated *k-UTDS graph*, if $\gamma_t(G) = \gamma_{\mathrm{gr}}^{\mathrm{t}}(G) = k$.

Clearly, given a graph G with a twin vertex x, G is a k-UDS graph if and only if $G - x$ is a k-UDS graph. Similarly, if x is an open twin vertex of G, then G is a k-UTDS graph if and only if $G - x$ is a k-UTDS graph. Hence to characterize these two classes of graphs, one can consider only (open) twin-free graphs.

In the following result, we characterize connected, twin-free, k-UDS graphs for all k.

Theorem 4.20 [19, Theorem 3.6] and [58, Theorem 2.5] *Let k be a positive integer. If G is a connected, twin-free, k-UDS graph, then either k = 1 and G = K_1, or k = 2 and G is a cocktail-party graph.*

The statement of Theorem 4.20 when $k \leq 3$ was proved in the seminal paper [19], while the general case was later resolved by Erey [58]. It is clear that the disjoint union of uniform dominating sequence graphs is again a uniform dominating sequence graph, hence one can build (disconnected) k-UDS graphs for arbitrarily large k.

While the structure of k-UDS graphs is completely understood, this is not the case with k-UTDS graphs as soon as $k > 3$. We start the discussion with a characterization of open twin-free 2-UTDS graphs.

Theorem 4.21 [21, Theorem 4.4] *If G is an isolate-free graph, then G is an open twin-free, 2-UTDS graph if and only if G is a complete graph.*

Proof. If $G = K_n$, then clearly G is an open twin-free graph with $\gamma_{gr}^t(G) = 2 = \gamma_t(G)$.

For the converse, suppose that x and y are distinct vertices in an isolate-free 2-UTDS graph G, which has no open twins. (In particular, $\gamma_{gr}^t(G) = 2$.) Suppose that x and y are not adjacent. Since x and y are not open twins, we may assume without loss of generality that there exists a vertex u, which has x as a neighbor, and u and y are not adjacent. Note that the sequence (x, u) is an open neighborhood sequence, and $\widehat{(x, u)}$ is not a total dominating set; hence, it can be extended to an open neighborhood sequence of length at least 3, which gives the contradiction $\gamma_{gr}^t(G) \geq 3$. This implies that x and y are adjacent, and hence G is a complete graph.   □

By Theorem 4.6, there are clearly no 3-UTDS graphs, while Gologranc, Jakovac, Kos, and Marc [54] characterized the 4-UTDS graphs among bipartite graphs as follows. (Notice that the resulting class of graphs is in a sense similar to the cocktail-party graphs that appear in Theorem 4.20.)

Theorem 4.22 [54, Theorem 3.1] *Let G be a bipartite, open twin-free graph with no isolated vertices. The graph G is a 4-UTDS if and only if $G = K_{n,n} - M$, where M is a perfect matching in $K_{n,n}$ and $n \geq 2$.*

It was conjectured that there are no non-bipartite 4-UTDS graphs [54, Conjecture 3.2]. However, this was disproved by Bahadır, Gözüpek, and Doğan [5], who Gözüpek noticed that the line graph $L(K_6)$ of the complete graph of order 6 is also a 4-UTDS graph. In addition, they proved the following result concerning odd values k in k-UTDS graphs.

Theorem 4.23 [5, Theorem 2.3] *There does not exist a k-UTDS graph if k is an odd positive integer.*

In particular, there are no 5-UTDS graphs; hence, the next case to consider are the 6-UTDS graphs. The situation there is even more interesting than with 4-UTDS graphs. An intriguing connection between the bipartite, regular 6-UTDS graphs, and some well-known concepts of design theory appears in the following result.

Theorem 4.24 [54, Theorem 4.7] *Let n and k be positive integers. There exists an $(n - k)$-regular bipartite, open twin-free, 6-UTDS graph G of order $2n$ if and only if $n = k^2 - k + 1$ and there exists a projective plane of order $k - 1$.*

There is an infinite family of graphs that appear in Theorem 4.24, since it is known that projective planes exist for each $k - 1$, which is a power of a prime number.

The authors of [5] made further progress in understanding the k-UTDS graphs. They proved that for every positive even integer k, a connected open twin-free k-UTDS graph is always regular [5, Theorem 4.2], and that there exist no connected chordal k-UTDS graphs when $k \geq 4$ [5, Theorem 5.2]. By using the direct product with the graph K_2, Bahadır et al. [5] proved the following result.

Theorem 4.25 [5, Theorem 3.4] *If G is a connected, non-bipartite k-UTDS graph, then $G \times K_2$ is a connected $(2k)$-UTDS graph.*

This result implies the existence of a connected 8-UTDS graph, namely, the graph $L(K_6) \times K_2$. The authors of [5] believe that there exist connected k-UTDS graphs for every even $k \geq 10$; however, this remains an open problem.

4.5   Z-Grundy Domination and Zero Forcing

In view of the definition of closed and open neighborhood sequences, the following definition is natural; see [13]. Let G be a graph without isolated vertices. A sequence $S = (v_1, \ldots, v_k)$ of vertices in G is a *Z-sequence* if, for each $i \in [k]$,

$$N(v_i) - \bigcup_{j=1}^{i-1} N[v_j] \neq \emptyset, \tag{4.6}$$

and the set $\widehat{S}$ is then a *Z-set*. Also, we say that v_i *Z-footprints* the vertices in $N(v_i) - \bigcup_{j=1}^{i-1} N[v_j]$. The *Z-Grundy domination number*, $\gamma_{gr}^Z(G)$, of G is the maximum length of a Z-sequence. A Z-sequence is also a closed neighborhood sequence. More precisely, in the condition required for a Z-sequence, a vertex v_i must dominate a neighbor that has not been dominated by $\{v_1, \ldots, v_{i-1}\}$, while in a closed neighborhood sequence v_i may also be the only vertex of its closed neighborhood that has not been dominated by $\{v_1, \ldots, v_{i-1}\}$. It follows from the definitions that $\gamma(G) \leq \gamma_{gr}^Z(G) \leq \gamma_{gr}(G)$.

From Proposition 4.1, we infer the following general upper bound for the Z-Grundy domination number (the sharpness of the bound can be demonstrated by paths and complete graphs).

Corollary 4.26 *If G is a graph with no isolated vertices, $\gamma_{gr}^Z(G) \leq n(G) - \delta(G)$.*

Kos [90] studied the effect of vertex or edge removal on the Z-Grundy domination number of a graph. All the bounds in the following two theorems were shown to be sharp; see [90] for details.

Theorem 4.27 [90, Theorem 3.23] *If G is a graph and $e \in E(G)$ such that $G - e$ has no isolated vertices, then*

$$\gamma_{gr}^Z(G) - 1 \leq \gamma_{gr}^Z(G - e) \leq \gamma_{gr}^Z(G) + 1.$$

Theorem 4.28 [90, Theorem 3.28] *If G is a graph with no isolated vertices and $v \in V(G)$ such that $G - v$ has no isolated vertices, then*

$$\gamma_{gr}^Z(G) - 2 \leq \gamma_{gr}^Z(G - v) \leq \gamma_{gr}^Z(G).$$

The main motivation for introducing the Z-Grundy domination number was its connection to the zero forcing number that has been studied earlier in graph theory and was also related to some concepts in linear algebra. (This is also the reason for using "Z" in the name of the concept.) Let us present these concepts.

Let G be a graph and let $A(G)$ be its adjacency matrix. Denote by $S(G)$ the set of all symmetric matrices in which ones from $A(G)$ are replaced by arbitrary non-zero real numbers, zeros outside the diagonal of $A(G)$ stay zeros, and zeros from the diagonal may be replaced by arbitrary real numbers. The minimum rank of a matrix in $S(G)$ is the *minimum rank*, $\mathrm{mr}(G)$, of G. The concept has been intensively studied by the linear algebra community, and in 2007 a close connection to the zero-forcing number was established [2].

Let $V(G)$ be partitioned into two sets, B and W, called blue and white vertices, respectively. A *color change operation* consists of choosing a vertex $x \in B$ that has exactly one white neighbor y, and changing the color of y to blue. If B is initially chosen in such a way that repeating the color change operation results in all vertices of W being colored blue, then B is called a *zero-forcing set*. The *zero-forcing number* of G, $Z(G)$, is the smallest cardinality of such a set B. The zero-forcing number can be used to bound the minimum rank of a graph G as follows: $n(G) - Z(G) \leq \mathrm{mr}(G)$; see [2]. This lower bound for the minimum rank of a graph can be expressed in terms of the Z-Grundy domination number, due to the following connection with the zero forcing number proved in [13] by Brešar, Bujtás, Gologranc, Klavžar, Košmrlj, Patkós, Tuza, and Vizer.

Theorem 4.29 [13, Theorem 2.2] *If G is a graph without isolated vertices, then*

$$\gamma_{gr}^Z(G) + Z(G) = n(G).$$

Moreover, the complement of a (minimum) zero forcing set of G is a (maximum) Z-set of G and vice versa.

Proof. Without loss of generality we may assume that G is connected.

Note that to prove Theorem 4.29, it suffices to prove that the complement of a zero forcing set of G is a Z-set and that the complement of a Z-set is a zero forcing set.

Let X be a zero forcing set of G, let $Y = V(G) - X$ and $|Y| = k$. Denote by x_i the vertex in G on which a color change operation is performed in the ith step; that is, in the ith step, x_i is colored blue and has exactly one white neighbor, say y_i, after which the color of y_i is changed to blue. Note that $x_1 \in X$ and $y_1 \in Y$. To prove that Y is a Z-set, we claim that $(y_k, \ldots, y_1)$ is a Z-sequence. Note that $x_k \neq y_k$. Since y_k Z-footprints all its neighbors, y_k Z-footprints x_k. Let $i \in [k-1]$. The vertex y_i Z-footprints x_i, because when the color of y_i was changed to blue, y_i was the only white neighbor of the blue vertex x_i.

The proof that the complement of a Z-set is a zero forcing set follows by a similar argument.   □

As mentioned above, by Theorem 4.29 we get $\gamma_{\mathrm{gr}}^Z(G) \leq \mathrm{mr}(G)$. In addition, combining Theorem 4.29 with the inequality $\gamma_{\mathrm{gr}}^Z(G) \leq \gamma_{\mathrm{gr}}(G)$, one also infers the inequality

$$\gamma_{\mathrm{gr}}(G) + Z(G) \geq n(G),$$

which holds for every graph G with no isolated vertices.

Additional two types of color change operations that yield another two versions of zero-forcing sets were considered in [2, 78]. The corresponding zero-forcing-type numbers are denoted by $Z_{\ell}(G)$ and $Z_-(G)$. In addition, there are two types of minimum rank parameters that correspond to these two zero forcing numbers, and are denoted by $\mathrm{mr}_{\ell}(G)$ and $\mathrm{mr}_0(G)$, respectively. The former is defined as the minimum rank of the matrices in $S(G)$ all of whose diagonal elements are non-zero, while in the latter the minimum rank is taken over all matrices in $S(G)$ all of whose diagonal elements are zero.

Lin [94, Theorem 2.2] proved that

$$\gamma_{\mathrm{gr}}(G) + Z_{\ell}(G) = n(G) \text{ and } \gamma_{\mathrm{gr}}^{\mathrm{t}}(G) + Z_-(G) = n(G)$$

holds for all graphs G. As an almost immediate consequence, the following connections of two types of Grundy domination numbers with the corresponding minimum rank parameters were established: $\gamma_{\mathrm{gr}}(G) \leq \mathrm{mr}_{\ell}(G)$ and $\gamma_{\mathrm{gr}}^{\mathrm{t}}(G) \leq \mathrm{mr}_0(G)$; see [94, Theorem 3.1]. In addition, by using algebraic properties of certain symmetric matrices, Lin proved in [94, Propositions 3.4 and 3.3] that $\mathrm{mr}_{\ell}(G) \leq \theta_e(G)$ and $\mathrm{mr}_0(G) \leq 2\beta(G)$. The former implies Proposition 4.2, while the latter implies the bound $\gamma_{\mathrm{gr}}^{\mathrm{t}}(G) \leq 2\beta(G)$, which was mentioned in Section 4.2.

4.6   L-Grundy Domination

Note that the conditions in (4.1), (4.4), and (4.6) that define closed neighborhood sequences, open neighborhood sequences, and Z-sequences, respectively, comprise three of the four possible ways in which the new vertex and the previously chosen vertices appear with their closed or their open neighborhoods. The missing one is now used to define L-Grundy domination (the letter L stands for the longest, since for any graph the L-Grundy sequence turns out to be at least as long as the other three types of sequences). This concept was introduced in [13].

Given a graph G, a sequence $S = (v_1, \ldots, v_k)$ of **distinct** vertices from G is called an *L-sequence* if, for each $i \in [k]$,

$$N[v_i] - \bigcup_{j=1}^{i-1} N(v_j) \neq \emptyset. \tag{4.7}$$

For an L-sequence S, the corresponding set $\widehat{S}$ of vertices is an *L-set*. Note that unless one requires that the vertices of an L-sequence are distinct, a repetition of any vertex in G complies with the condition in (4.7). The *L-Grundy domination number*, $\gamma_{gr}^L(G)$, of the graph G is the maximum length of an L-sequence.

By definitions of the involved invariants, we immediately infer $\gamma_{gr}(G) \leq \gamma_{gr}^L(G)$ and $\gamma_{gr}^t(G) \leq \gamma_{gr}^L(G)$. Combining the latter inequality with Theorem 4.8, we get a large class of graphs for which $\gamma_{gr}^L(G) = n(G)$. However, there are more such graphs, and the characterization of them is still unknown.

Problem 4.30 [13] Characterize the class of graphs G with $\gamma_{gr}^L(G) = n(G)$.

Brešar, Gologranc, Henning, and Kos [17] gave a partial answer to the above problem with the following result.

Theorem 4.31 [17, Theorem 4.1] *If G is a graph with $\gamma_{gr}^L(G) = n(G)$, then $\delta(G) \leq 1$.*

In addition, they proved that $\gamma_{gr}^L(T) = n(T)$ for every forest T, and designed an algorithm that determines an L-sequence of length $n(T)$ in a forest T [17, Section 5].

Brešar et al. also established the following lower bound on the L-Grundy domination number of a k-regular connected graph that is different from $K_{k,k}$ and K_{k+1}.

Theorem 4.32 [17, Theorem 3.4] *If $k \geq 3$ and G is a k-regular connected graph of order n different from $K_{k,k}$ and K_{k+1}, then*

$$\gamma_{gr}^L(G) \geq \frac{2(k-2)n + (4-k)\alpha(G)}{(k-1)^2}.$$

In particular, if G is a cubic graph different from $K_{3,3}$ and K_4, Theorem 4.32 gives $\gamma_{\mathrm{gr}}^{\mathrm{L}}(G) \geq \frac{1}{2}n(G) + \frac{1}{4}\alpha(G)$.

The effect of removing a vertex or an edge on the L-Grundy domination number of a graph was studied by Kos [90]; it was also shown that the bounds are sharp.

Theorem 4.33 [90, Theorem 3.25] *If G is a graph and $e \in E(G)$, then*

$$\gamma_{\mathrm{gr}}^{\mathrm{L}}(G) - 1 \leq \gamma_{\mathrm{gr}}^{\mathrm{L}}(G - e) \leq \gamma_{\mathrm{gr}}^{\mathrm{L}}(G) + 2.$$

Theorem 4.34 [90, Theorem 3.33] *If G is a graph and $v \in V(G)$, then*

$$\gamma_{\mathrm{gr}}^{\mathrm{L}}(G) - 2 \leq \gamma_{\mathrm{gr}}^{\mathrm{L}}(G - v) \leq \gamma_{\mathrm{gr}}^{\mathrm{L}}(G).$$

The authors of [17] conjectured that an upper bound for the L-Grundy domination number analogous to that in Proposition 4.7 for the Grundy total domination number holds.

Conjecture 4.35 [17, Conjecture 4.3] If G is a graph, then $\gamma_{\mathrm{gr}}^{\mathrm{L}}(G) \leq n(G) - \delta(G) + 1$.

The conjecture was proved to be true when G is a graph with $\delta(G) \leq 2$; see [17, Corollary 4.2].

Motivated by the connections between different types of Grundy domination numbers, zero-forcing numbers, and minimum rank parameters, Lin [94] introduced the *L-zero-forcing number*, $Z_L(G)$, of a graph G, and a new minimum rank parameter denoted by $\mathrm{mr}_L(G)$. Similar results to those at the end of Section 4.5 can be obtained for these invariants. In particular, in every graph G, $\gamma_{\mathrm{gr}}^{\mathrm{L}}(G) \leq \mathrm{mr}_L(G)$; see [94] for definitions and for more details.

4.7 Comparison of the Four Grundy Domination Invariants

The following theorem summarizes the relations between all pairs of the four Grundy domination invariants.

Theorem 4.36 *The bounds presented in the following table are correct and sharp.*

	$\leq$	1 γ_{gr}^{Z}	2 γ_{gr}	3 γ_{gr}^{t}	4 $\gamma_{\mathrm{gr}}^{\mathrm{L}}$
1	γ_{gr}^{Z}	$=$	γ_{gr}	γ_{gr}^{t}	$\gamma_{\mathrm{gr}}^{\mathrm{L}} - 1$
2	γ_{gr}	$\nexists f$	$=$	$\nexists f$	$\gamma_{\mathrm{gr}}^{\mathrm{L}} - 1$
3	γ_{gr}^{t}	$2\gamma_{\mathrm{gr}}^{Z}$	$2\gamma_{\mathrm{gr}}$	$=$	$\gamma_{\mathrm{gr}}^{\mathrm{L}}$
4	$\gamma_{\mathrm{gr}}^{\mathrm{L}}$	$\nexists f$	$2\gamma_{\mathrm{gr}}$	$\nexists f$	$=$

The entry $(i, j) \in [4] \times [4]$ representing the ordered pair of invariants $(\gamma_{gr}^*, \gamma_{gr}^+)$, both of which are from $\{\gamma_{gr}, \gamma_{gr}^t, \gamma_{gr}^Z, \gamma_{gr}^L\}$, has the following meaning when $i \neq j$. If the entry equals $\nexists f$, then there exists no function f such that $\gamma_{gr}^*(G) \leq f(\gamma_{gr}^+(G))$ holds for every graph G that is well defined for both invariants γ_{gr}^* and γ_{gr}^+. Otherwise, it indicates that for every graph G that is well defined for both invariants, we have $\gamma_{gr}^*(G) \leq f(\gamma_{gr}^+(G))$, where f is a function of γ_{gr}^+ given in entry (i, j).

The entry $(3, 2)$ is proven in [21, Theorem 7.1], while the non-existence of a function f such that $\gamma_{gr}(G) \leq f(\gamma_{gr}^t(G))$ would hold for all graphs G is demonstrated by the class of stars, $K_{1,n}$. In fact, the class of stars can be used to verify all the entries $(2, 1)$, $(2, 3)$, $(4, 1)$, and $(4, 3)$. Entries $(1, 2)$, $(1, 3)$, and $(3, 4)$ are clear. The entry $(2, 4)$ was proven in [13, Proposition 3.1], while $(1, 4)$ follows from combining the entries $(1, 2)$ and $(2, 4)$. The entry $(3, 1)$ was shown in [13, Proposition 3.3], while the entry $(4, 2)$ was proven in [13, Proposition 3.4].

To see that the bounds in the table of Theorem 4.36 are sharp, take the following examples. The entries $(1, 2)$ and $(1, 4)$ are sharp for paths P_n; entries $(1, 3)$ and $(2, 4)$ are sharp for stars. The entry $(3, 4)$ is covered by Theorem 4.8, which characterizes the graphs G with $\gamma_{gr}^t(G) = n(G)$. Finally, the entries $(3, 1)$, $(3, 2)$, and $(4, 2)$ are sharp because of the family of graphs G_k^n obtained from k copies of the complete graph K_n, $n \geq 3$, by choosing a vertex x_i from the ith copy, for each $i \in [k]$, and then identifying the vertices $x_1, \ldots, x_k$. In the resulting graph, $\gamma_{gr}^t(G_k^n) = 2k = \gamma_{gr}^L(G_k^n)$, while $\gamma_{gr}^Z(G_k^n) = k = \gamma_{gr}(G_k^n)$.

4.8 Complexity of the Grundy Domination Concepts

In this section, we consider the decision versions of the four Grundy domination invariants. We start with the Grundy domination number, where the decision problem is the following:

GRUNDY DOMINATION NUMBER

Input: A graph G, and an integer k.
Question: Is $\gamma_{gr}(G) \geq k$?

In the seminal paper on Grundy dominating sequences [19], the NP-completeness of GRUNDY DOMINATION NUMBER was established by using Theorem 4.18 and a (non-trivial) reduction from GRUNDY COVERING NUMBER IN HYPERGRAPHS.

Theorem 4.37 [19, Theorem 4.1] GRUNDY DOMINATION NUMBER *is NP-complete even when restricted to chordal graphs.*

GRUNDY TOTAL DOMINATION NUMBER

Input: A graph G with no isolated vertices, and an integer k.
Question: Is $\gamma_{gr}^t(G) \geq k$?

The complexity of GRUNDY TOTAL DOMINATION NUMBER was first studied by Brešar, Henning, and Rall in [21], where NP-completeness was established in the class of bipartite graphs. The same complexity was proved for split graphs by Brešar, Kos, Nasini, and Torres in [26].

Theorem 4.38 [21, Theorem 4.1] and [26, Theorem 6.1] GRUNDY TOTAL DOMI-NATION NUMBER *is NP-complete even when restricted to bipartite or split graphs.*

The proof for bipartite graphs is obtained by the following reduction. Let $H = (V, F)$ be a hypergraph. The *incidence graph* of H, denoted by B_H, is the bipartite graph with V and F as partite sets, and $x \in V$ is adjacent to $e \in F$ if and only if $x \in e$. We claim that

$$\gamma_{gr}^t(B_H) = 2\rho_{gr}(H).$$

Indeed, if S is a (longest) open neighborhood sequence in B_H, then $S \cap F$ coincides with a (longest) covering sequence in H, hence its length is $\rho_{gr}(H)$. Similarly, $S \cap V$ coincides with a (longest) transversal sequence in H of length $\tau_{gr}(H)$. This gives $\gamma_{gr}^t(B_H) = \rho_{gr}(H) + \tau_{gr}(H)$, and Proposition 4.19 implies the claimed equality. Since GRUNDY COVERING NUMBER IN HYPERGRAPHS is NP-complete, we infer the same for the GRUNDY TOTAL DOMINATION NUMBER in bipartite graphs.

The study of zero-forcing began much earlier than that of Grundy domination. Hence it is not surprising that the complexity of the decision version of the zero-forcing number was established already in 2009 by Aazami; see [1]. The corresponding problem is to verify whether $Z(G) \leq k$ holds for a given graph G and a positive integer k. In the context of Z-Grundy domination, we are considering its dual problem, which is as follows.

Z-GRUNDY DOMINATION NUMBER

Input: A graph G with no isolated vertices, and an integer k.
Question: Is $\gamma_{gr}^Z(G) \geq k$?

By Theorem 4.29, we infer that the complexity of Z-GRUNDY DOMINATION NUMBER is the same as the complexity of the decision version of the zero-forcing number, which was proved to be NP-complete [1]. Hence, we derive the following result.

Corollary 4.39 Z-GRUNDY DOMINATION NUMBER *is NP-complete.*

L-Grundy domination was introduced in [13], where its computational complexity was also considered. Notably, the problem

---

L-GRUNDY DOMINATION NUMBER

Input: A graph G, and an integer k.
Question: Is $\gamma_{gr}^{L}(G) \geq k$?

---

was shown to be NP-complete as stated by the following result.

Theorem 4.40 [13, Theorem 4.1] and [90, Theorem 3.48] L-GRUNDY DOMINA-TION NUMBER *is NP-complete even when restricted to bipartite graphs or split graphs.*

4.9 Classes of Graphs: Algorithms and Exact Values

Brešar, Gologranc, Milanič, Rall, and Rizzi [19] studied the Grundy domination number in trees. They presented a simple, linear algorithm to determine a dominating sequence in a tree T of length $n(T) - |ES(T)| + 1$, where $ES(T)$ is the set of all support vertices in T that do not lie on a path between two other support vertices. While this may not be a Grundy dominating sequence, a dynamic programming algorithm to determine a Grundy dominating sequence in linear time was given in [19, Algorithm 2]. For the Grundy total domination number in a forest F, Brešar, Kos, Nasini, and Torres [26, Theorem 5.1] proved that $\gamma_{gr}^{t}(F) = 2\beta(F)$, and presented a linear algorithm to find a Grundy total dominating sequence in a tree T.

It was shown by Kos [90] that $\gamma_{gr}^{L}(F) = n(F)$ for any forest F. In addition, the order of vertices of F, which is an L-dominating sequence of F, was given in [90, Algorithm 2]. The Z-Grundy domination number in trees was implicitly considered in the context of minimum rank. Notably, the result [59, Algorithm 2.5] by Fallat and Hogben gives a linear algorithm to compute $mr(T)$. By [59, Theorem 2.2], $mr(T) = n(T) - P(T)$ in an arbitrary tree, where $P(T)$ is the *path cover number*, that is, the smallest number of paths into which $V(T)$ can be partitioned. Now, it was proved by Hogben in [78, Corollary 3.6] that $P(T) = Z(T)$ in any tree T. We thus derive that there is an efficient algorithm to determine $Z(T)$, and thus also $\gamma_{gr}^{Z}(T)$, in any tree T. In [103], Row extended this algorithm to all unicyclic graphs.

Recall that GRUNDY TOTAL DOMINATION NUMBER and L-GRUNDY DOMI-NATION NUMBER are NP-complete problems in split graphs. In contrast, one can efficiently compute $\gamma_{gr}(G)$ for a split graph G with a partition (K, I), where I is a maximum independent set and K induces a complete subgraph. It was shown in [19, Theorem 2.6] that $\gamma_{gr}(G) = \alpha(G)$ if every two vertices of K have a common neighbor in I, and $\gamma_{gr}(G) = \alpha(G)+1$ otherwise. Since the independence number of a split graph can be computed efficiently, this is true also for the Grundy domination number. To the best of our knowledge, the complexity of Z-GRUNDY DOMINATION NUMBER in split graphs is still unknown.

As shown in [19], $\gamma_{gr}(G) = \alpha(G)$ holds in every cograph G. Since the independence number of a cograph is efficiently computable, this is so also for the Grundy domination number. In addition, Kos [90, Theorem 3.70] showed that the computation of all four Grundy domination invariants is linear in the class of cographs. Cographs have been generalized into different graph classes, which have the property that there are not many induced subgraphs isomorphic to P_4. Nasini and Torres [98] presented polynomial algorithms for computing the Grundy domination number in several such families of graphs with few P_4s. In addition, they considered the so-called X-join operation on graphs and provided algorithmic results related to Grundy domination using this operation [98]. Brešar, Kos, Nasini, and Torres [26, Theorem 5.1] provided a linear algorithm for the Grundy total domination number in a class of graphs with few induced P_4s and in distance-hereditary graphs.

Recall that an *interval graph* is the intersection graph on a finite collection of intervals on the real line. A linear algorithm to find a Grundy dominating sequence of an interval graph was presented by Brešar, Gologranc, and Kos [18, Algorithm 1]. When an interval graph is represented by the increasing sequence of the endpoints of the real line intervals, the Grundy domination number coincides with the number of consecutive subsequences of length 2, in which the first endpoint is a left endpoint and the second endpoint is a right endpoint. (The preprocessing of an interval graph to be represented by such a sequence can be done in $O(n \log n + m)$ time.) In the same paper [18, Theorem 5], the authors gave an efficient algorithm to determine the Grundy domination number of a so-called Sierpiński graph; it is given by a closed formula. In [13, Proposition 2.6], it was proved that the same algorithm and formula works for Z-Grundy domination.

Grundy domination number in the four standard graph products was considered in [12]. Beside some general bounds that were obtained, the emphasis was on obtaining exact results for products of paths and/or cycles. In particular, for the Cartesian product and the lexicographic product of paths and/or cycles, exact values of the Grundy domination numbers were found [12, Theorem 4 and Corollaries 11-14]. (For the strong and the direct product of paths and/or cycles, either exact values or lower and upper bounds were established [12].) The Grundy total domination number of the four graph products was addressed in [10]. In particular, exact values for the direct and the lexicographic product of paths and/or cycles were found, and lower bounds were proved for the direct and the lexicographic product of two arbitrary graphs. For the strong and the Cartesian product, the authors established lower and upper bounds when each factor is either a cycle or a path.

The following compelling conjecture was made in [12]. (It is straightforward to see that $\gamma_{gr}(G \boxtimes H) \geq \gamma_{gr}(G)\gamma_{gr}(H)$ holds for any two graphs G and H.)

Conjecture 4.41 [12, Conjecture 22] If G and H are graphs, then

$$\gamma_{gr}(G \boxtimes H) = \gamma_{gr}(G)\gamma_{gr}(H).$$

Interestingly, a close connection between the Grundy domination number in strong products and the Grundy total domination number in direct products was found in [10]. In particular, the following conjecture is equivalent to Conjecture 4.41 as shown in [10], where again the inequality $\gamma_{gr}^t(G \times H) \geq \gamma_{gr}^t(G)\gamma_{gr}^t(H)$ is straightforward.

Conjecture 4.42 [10, Conjecture 2.2] If G and H are graphs, then

$$\gamma_{gr}^t(G \times H) = \gamma_{gr}^t(G)\gamma_{gr}^t(H).$$

The equivalence of the above two conjectures follows from their equivalence with another conjecture on a hypergraph product structure concerning the Grundy covering number; see [10]. Conjecture 4.42 was confirmed when one of the factors, say G, satisfies $\gamma_{gr}^t(G) = bc(G)$ [10] (recall that $bc(G)$ is the smallest size of a covering of the edges of G with complete bipartite graphs). In particular, this holds for trees. On the other hand Conjecture, 4.41 is true if one of the factors is a caterpillar [12].

The authors in [10] also considered the Z-Grundy domination number and the L-Grundy domination number of the lexicographic product of two arbitrary graphs.

Given integers n and r, where $n \geq 2r \geq 2$, the *Kneser graph* $K(n, r)$ is the graph whose vertices are all the r-subsets of $[n]$, and two r-sets are adjacent if they are disjoint. The four types of Grundy domination numbers in Kneser graphs were considered by Brešar, Kos, and Torres in [27]. A closed formula was obtained for the Grundy total domination number and reads:

$$\gamma_{gr}^t(K(n, r)) = \binom{2r}{r}.$$

The same value is obtained for the Z-Grundy domination number if $n \geq 3r + 1$. When $2r \leq n \leq 3r$, both lower and upper bounds were found for $\gamma_{gr}^Z(K(n, r))$. On the other hand, it was proved that $\gamma_{gr}(G) = \alpha(K(n, r)) = \binom{n-1}{r-1}$ as soon as n is large enough compared to r. For instance, the Grundy domination number of the Petersen graph, $K(5, 2)$, equals 5, while $\gamma_{gr}(K(n, 2)) = n - 1 = \alpha(K(n, 2))$ for $n > 5$. Lower and upper bounds for $\gamma_{gr}^L(K(n, r))$ were also obtained in [27].

We have mentioned a large majority of the known algorithmic and exact results on the four Grundy domination invariants. In addition to these, several results concerning the zero-forcing number of special classes of graphs can be found in the literature, many of which are expressed in the language and from the perspective of linear algebra. Let us only mention that a number of exact values or bounds for some classes of graphs that are well known in graph theory were presented in the seminal paper [2], in which zero-forcing was introduced. In particular, several results concerning the Z-Grundy domination number in various classes of graphs can be derived from [2] by using Theorem 4.29. For example, $\gamma_{gr}^Z(P_s \square P_t) = s(t - 1)$ for $2 \leq s \leq t$, $\gamma_{gr}^Z(P_s \boxtimes P_t) = (s - 1)(t - 1)$, $\gamma_{gr}^Z(Q_n) = 2^{n-1}$, where Q_n is the n-cube (that is, the Cartesian product of n copies of K_2).

Chapter 5
Related Games on Graphs and Hypergraphs

A number of variations of the domination game and the total domination game have been introduced. Some appeared earlier than 2010, when the seminal paper on the domination game was published. For some reason these games had not received a lot of attention, and the introduction of the new domination game(s) brought them back to the light. In addition, a number of new games appeared in the last decade, which were triggered by the success of the domination game.

One of the first domination-type games that was introduced after the domination game was established is the disjoint domination game, which we present in Section 5.1. In this game, proposed by Bujtás and Tuza, one of the players wants to obtain a partition of the vertex set into two dominating sets while the other player wants to prevent this to happen. Recently, the same two authors introduced the fractional version of the domination game, which we present in Section 5.2. In Section 5.3, we present the connected domination game, introduced by Borowiecki, Fiedorowicz, and Sidorowicz, which is a version of the domination game that corresponds to connected domination. Next, we follow with game versions of the problems related to Z-sequences and L-sequences, as introduced by Brešar, Bujtás, Golografnc, Klavžar, Košmrlj, Marc, Patkós, Tuza, and Vizer. Sections 5.5 and 5.6 are concerned with games played on hypergraphs, all of which have a similar set of rules as the domination game. The first one is the transversal game proposed by Bujtás, Henning, and Tuza, and the second one, due to Bujtás, Patkós, Tuza, and Vizer, is the (total) domination game. In Section 5.7 we present several other games related to the domination game(s). We begin with the recently introduced Maker-Breaker domination game. Then we present two games introduced by Phillips and Slater back in 2001. First, the enclaveless game is inspired by the so-called enclaveless number of a graph, which is in a sense dual to the domination number. Second, they introduced the independent domination game, which is a variation of the domination game in which players are building an independent dominating set. Haynes and Henning recently proposed a variation of the domination game with

B. Brešar et al., *Domination Games Played on Graphs*, SpringerBriefs in Mathematics, https://doi.org/10.1007/978-3-030-69087-8_5

players building a paired-dominating set. Finally, we present the oriented version of the domination game, introduced back in 2002 by Alon, Balogh, Bollobás, and Szabó.

5.1 Disjoint Domination Game

In 2016 Bujtás and Tuza [45] introduced and studied the disjoint domination game. It is well known that every graph without isolated vertices contains two disjoint dominating sets. In this combinatorial game introduced by Bujtás and Tuza, one of the two players aims at constructing two disjoint dominating sets, while the other player wants to avoid this.

More formally, the *disjoint domination game* is a two-player game, where the players are named Dom (short for "Dominator") and Sepy (short for "Separator"). The game is played in an isolate-free graph G with a given color palette $C = \{p, b\} = \{$purple$,$ blue$\}$. At any stage of the game, V_p and V_b denote the set of vertices colored with p and b, respectively. For $c \in C$, we denote by $\bar{c}$ the complementary color for which $\{c, \bar{c}\} = C$ holds. Thus, $\bar{b} = p$ and $\bar{p} = b$. In the game, Dom and Sepy take turns choosing a vertex and assigning it with a color from C, where a vertex v and its coloring with a color $c \in C$ is *legal* (or *feasible*) in the game if and only if the following hold.

(i) The vertex v has not yet been chosen and assigned with a color, that is, $v \notin V_p \cup V_b$.
(ii) There exists a vertex $u \in N[v]$ which has not been dominated in color c, that is, $N[u] \cap V_c = \emptyset$.

Each player is required to select a vertex on his turn whenever a legal move is available. The game terminates when one of the following situations is reached.

$\langle s^* \rangle$ Some vertex has a monochromatic closed neighborhood.
$\langle d^* \rangle$ Both V_p and V_b are dominating sets.

The winner is Sepy if $\langle s^* \rangle$ is reached, and Dom wins if $\langle d^* \rangle$ is reached. Thus, Dom and Sepy have conflicting objectives, namely the aim of Dom is to obtain two disjoint dominating sets V_p and V_b at the end of the game, while the aim of Sepy is to prevent this situation from occurring. As observed in [45], as long as neither $\langle s^* \rangle$ nor $\langle d^* \rangle$ is reached, the next player has a feasible move, implying that the game always ends with a win for one of the players.

Consider the Sepy-start disjoint domination game played in a connected isolate-free graph G. After each move of Sepy, Dom applies what Bujtás and Tuza [45] coin the *opposite neighbor strategy*, abbreviated ONS. In this strategy, whenever Sepy selects a vertex v and colors it with color c, then Dom responds by choosing a vertex according to the following rules.

- (ONS1) Dom selects a neighbor u of v which can be colored with $\bar{c}$, if possible.
- (ONS2) If (ONS1) cannot be applied, then Dom selects a vertex u which can be colored with color $c' \in C$ and that has a neighbor of color $\overline{c'}$.

As shown in [45], the opposite neighbor strategy guarantees that after each move of Dom, every colored vertex has a neighbor assigned with the opposite color, and furthermore, the situation $\langle s^* \rangle$ cannot occur. Moreover, after each selection of Sepy, Dom can always apply ONS if the game is not yet over, implying that the game will terminate with situation $\langle d^* \rangle$. We state this formally as follows.

Theorem 5.1 [45, Theorem 3] *If G is a connected isolate-free graph, then Dom has a winning strategy for the Sepy-start disjoint domination game in G.*

In contrast, the situation for the Dom-start disjoint domination game is more complex. As shown in [45], some graphs admit a winning strategy for Sepy in the Dom-start game. For example, Sepy has a winning strategy in the Dom-start game when played on a cycle on at least eight vertices, or when played on a graph obtained from a graph with minimum degree at least 2 by subdividing every edge twice. However, some graphs admit a winning strategy for Dom in the Dom-start game. For example, Dom has a winning strategy in the Dom-start game when played on a path P_n or a complete graph K_n. More generally, we have the following result.

Theorem 5.2 [45, Theorem 8] *If G is a connected isolate-free graph, then Dom has a winning strategy for the Dom-start disjoint domination game if G contains two distinct vertices u and v such that $N[u] \subseteq N[v]$.*

As remarked in [45], it is not very well understood which properties of a graph ensure a winning strategy for Dom or Sepy in the Dom-start disjoint domination game. Further, the authors remark that there exist graphs such that the possibility of passing (where we relax the requirement of a player selecting a vertex on his turn whenever a legal move is available) does increase Sepy's chances to win.

By definition, we exclude the possibility of passing in the very first move, because it would immediately change the character of the game by switching between Dom-start and Sepy-start. Passing may or may not help a player.

Theorem 5.3 [45, Theorem 3′] *If Dom is allowed to pass but Sepy is not, then Dom has a winning strategy in the Sepy-start disjoint domination game on every graph. Also, Dom has a winning strategy in the Dom-start disjoint domination game on every graph containing at least one Dom-win component.*

On the other hand, passing has no benefit for Sepy if he begins the game.

Theorem 5.4 [45, Theorem 3″] *If G is connected, then Dom has a winning strategy for the Sepy-start disjoint domination game on G, even when Sepy is allowed to pass at any time except in the first move.*

A biased version of the disjoint domination game is also studied in [45], where here a player may make more than one move at a time. The disjoint domination game where Dom sequentially selects and colors exactly d vertices in each turn (except near the end of the game when only fewer than d possibilities remain), and Sepy colors at most s vertices per turn (and may pass if he wishes to do so), we denote by the $(d : s)$-game. Thus, the $(1 : 1)$-game when $d = s = 1$, is precisely the disjoint domination game where Sepy is allowed to pass. Bujtás and Tuza [45] show that any $d > 1$ yields substantial advantage for Dom.

Theorem 5.5 [45, Theorem 10] *Dom has a winning strategy in the $(d : 1)$-game on every graph, for every $d \geq 2$.*

In the *bicolored disjoint domination game* we restrict Dom and Sepy to have their private colors, that is, Dom may only use color p and Sepy may only use color b. In this game as before, the situation $\langle s^* \rangle$ may occur, and the game terminates. However, the meaning of $\langle d^* \rangle$ is modified slightly here as follows:

- $\langle d^{**} \rangle$: For a color $c \in C$, V_c dominates all vertices of G, and no vertex v has its closed neighborhood entirely contained in V_c.

Using four different strategies, depending on the structure of the graph, Dom can win the bicolored disjoint domination game on any isolate-free graph.

Theorem 5.6 [45, Theorem 12] *Dom can win the bicolored disjoint domination game on every isolate-free graph, both in the Dom-start and Sepy-start cases.*

5.2 Fractional Domination Game

Bujtás and Tuza [46] introduced the fractional domination game as follows. The game on a graph G is, of course, played by Dominator and Staller who take turns. In each move a player distributes weight 1 to the vertices of G, more precisely, an ith *move*, $i \geq 1$, is a sequence

$$(u_{i_1}, w_1), (u_{i_2}, w_2), \ldots \qquad (5.1)$$

of arbitrary length (possibly infinite), where $u_{i_1}, u_{i_2}, \ldots$ are vertices of G with repetitions allowed, and $w_1, w_2, \ldots$ are real numbers from the interval $(0, 1]$. The elements (u_{i_j}, w_j) of the sequence (5.1) are called the *submoves* of the move (5.1). As said, a player distributes weight 1 to the vertices, that is, $\sum_{k \geq 1} w_k = 1$ must hold with the exception of the last move for which this condition is relaxed to $\sum_{k \geq 1} w_k \leq 1$.

To describe when the move (5.1) is legal and to define the game fractional domination number, we introduce *partially dominating functions* d_i, $i \geq 0$, and *load functions* ℓ_i, $i \geq 0$. At the beginning of the game we set $d_0(u_j) = \ell_0(u_j) = 0$ for every $u_j \in V(G)$. After the ith move (5.1), $i \geq 1$, the partially dominating function d_i is defined with

$$d_i(u_j) = d_{i-1}(u_j) + \sum_{\substack{k \geq 1 \\ i_k = j}} w_k .$$

That is, to the previous value $d_{i-1}(u_j)$ we add the total weight assigned to u_j in the ith move. From the partially dominating function d_i, the load function ℓ_i is then computed as follows:

$$\ell_i(u_j) = \min\{1, \sum_{x \in N[u_j]} d_i(x)\}.$$

Now, the move (5.1) is *legal* if for every $k \geq 1$ there exists a vertex $x \in N[u_{i_k}]$ such that

$$\ell_{i-1}(x) + w_k + \sum_{\substack{x \in N[u_{i_s}] \\ s \in [k-1]}} w_s \leq 1.$$

The game is finished when the load function is identically equal to 1. If this happens in the qth move of the game $\mathcal{G}$, then the *value* of $\mathcal{G}$ is

$$|\mathcal{G}| = \sum_{u \in V(G)} d_q(u).$$

That is, $|\mathcal{G}| = (q-1) + \epsilon$ if q moves were made when the load function became identically equal to 1 and the last player to play distributed a weight of ϵ, for $0 \leq \epsilon \leq 1$. Dominator's goal is that $|\mathcal{G}|$ is as small as possible, and Staller has the opposite goal. The *game fractional domination number* $\gamma_g^*(G)$ is then

$$\gamma_g^*(G) = \inf\{a : \text{Dominator has a strategy which ensures } |\mathcal{G}| \leq a\}.$$

The game fractional domination number can also be introduced through the eyes of Staller. More precisely, in [46, Proposition 1] it is observed that

$$\gamma_g^*(G) = \sup\{b : \text{Staller has a strategy which ensures } |\mathcal{G}| \geq b\}.$$

The *Staller-start game fractional domination number* $\gamma_g^{*'}(G)$ is defined analogously.

As an example consider the fractional domination game played on C_5 with the vertices $u_1, \ldots, u_5$. Let the first move of Dominator be

$$d_1 = (u_1, 1/5), (u_2, 1/5), (u_3, 1/5), (u_4, 1/5), (u_5, 1/5),$$

see Figure 5.1, where the index i in $(\cdot)_i$ indicates the order of the submoves. After this move we have $d_1(u_i) = 1/5$ and $\ell_1(u_i) = 3/5$ for $i \in [5]$. Suppose that Staller replies with the move $s_1 = (u_1, 2/5), (u_5, 2/5), (u_2, 1/5)$, see Figure 5.1 again. After this move we have $\ell_2(u_i) = 1$ for $i \in \{1, 2, 4, 5\}$ and $\ell_2(u_3) = 4/5$. Dominator can now finish the game, say with the move $(u_3, 1/5)$ (see Figure 5.1 once more), so that the value of this game $\mathcal{G}$ is $|\mathcal{G}| = 11/5$.

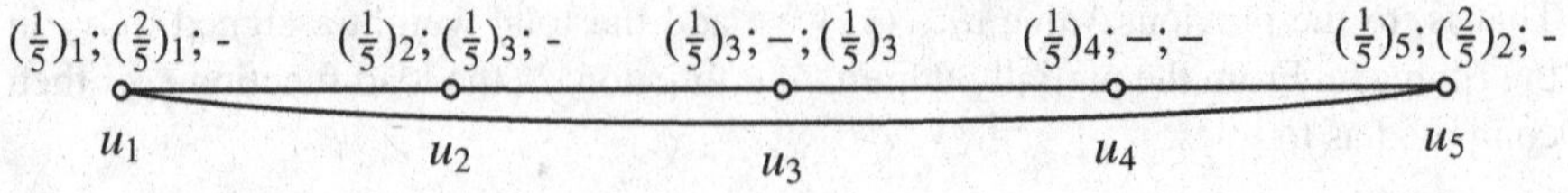

$(\frac{1}{5})_1; (\frac{2}{5})_1; -$ $\qquad$ $(\frac{1}{5})_2; (\frac{1}{5})_3; -$ $\qquad$ $(\frac{1}{5})_3; -; (\frac{1}{5})_3$ $\qquad$ $(\frac{1}{5})_4; -; -$ $\qquad$ $(\frac{1}{5})_5; (\frac{2}{5})_2; -$

u_1 $\qquad\qquad$ u_2 $\qquad\qquad$ u_3 $\qquad\qquad$ u_4 $\qquad\qquad$ u_5

Fig. 5.1 A fractional domination game played on C_5.

Suppose now that instead of the move s_1, Staller plays the move $s_1' = (u_5, 2/5)$, $(u_2, 1/5)$, $(u_1, 2/5)$, which is just a permutation of the submoves of s_1. This move, however, is not legal! Indeed, after the first two submoves are played, the load of every vertex from $N[u_1]$ will be strictly larger than $3/5$, hence when the submove $(u_1, 2/5)$ is played, the load of every vertex from $N[u_1]$ increases by less than the weight $2/5$ in the submove. This example shows that in the definition of a move we cannot replace the order of the submoves with the set of assignments to vertices.

In [46] it is actually proved that $\gamma_g^*(C_5) = 11/5$, the tricky part of the proof being a strategy of Staller which ensures that $\gamma_g^*(C_5) \geq 11/5$. Moreover, the values of $\gamma_g^*(P_n)$, $\gamma_g^*(C_n)$, $\gamma_g^{*\prime}(P_n)$, and $\gamma_g^{*\prime}(C_n)$ are determined for all $n \leq 6$. For $n \geq 7$ lower and upper bounds are proved which differ by a small additive constant.

A fine property of the domination game that extends to the fractional game is the Continuation Principle. If G is a graph and ℓ a load function, then let $\gamma_g^*(G|\ell)$ denote the *game fractional ℓ-domination number* which is defined analogously as $\gamma_g^*(G)$ with the sole exception that now the initial load function before the first move of Dominator is ℓ. The *Staller-start game fractional ℓ-domination number* $\gamma_g^{*\prime}(G|\ell)$ of G has an analogous meaning. Then Bujtás and Tuza proved the following [46, Theorem 4].

Theorem 5.7 (Fractional Continuation Principle) *If ℓ_1 and ℓ_2 are load functions on the graph G, where $\ell_1(u) \leq \ell_2(u)$ for every $u \in V(G)$, then $\gamma_g^*(G|\ell_1) \geq \gamma_g^*(G|\ell_2)$ and $\gamma_g^{*\prime}(G|\ell_1) \geq \gamma_g^{*\prime}(G|\ell_2)$.*

Just as for the standard game this implies the following important property.

Corollary 5.8 [46, Theorem 5] *If G is a graph, then $|\gamma_g^*(G) - \gamma_g^{*\prime}(G)| \leq 1$.*

Proof. Consider first the D-game and let ℓ be the load function after an optimal first move of Dominator. Since Dominator assigns total weight 1, we have $\gamma_g^*(G) = 1 + \gamma_g^{*\prime}(G|\ell)$. Since $\gamma_g^{*\prime}(G) \geq \gamma_g^{*\prime}(G|\ell)$ holds by the Fractional Continuation Principle, we get $\gamma_g^*(G) \leq 1 + \gamma_g^{*\prime}(G)$.

In the S-game, by a similar argument as above we get $\gamma_g^{*\prime}(G) \leq 1 + \gamma_g^*(G)$. Putting both inequalities together we arrive at $\gamma_g^*(G) - 1 \leq \gamma_g^{*\prime}(G) \leq 1 + \gamma_g^*(G)$, and the assertion follows. $\qquad\qquad\qquad\qquad\qquad\qquad\qquad\qquad\qquad\qquad\square$

Applying Alon's result from [3], we get that the upper bound $n(G)\frac{1+\ln(\delta(G)+1)}{\delta(G)+1}$ for the domination number of graphs of minimum degree δ is asymptotically tight as $\delta \to \infty$. From this fact Bujtás and Tuza derived the following result.

Proposition 5.9 [46, Proposition 3] *There does not exist a universal constant C such that for every graph G the following holds: $\gamma_g(G) \leq C \cdot \gamma_g^*(G)$.*

5.3 Connected Domination Game

The *connected domination game* on a graph G is played by Dominator and Staller according to the rules of the standard domination game with the additional requirement that at each stage of the game the selected vertices induce a connected subgraph of G. As usual, if both players play optimally, then the number of moves played is unique; in the D-game it is called the *game connected domination number* $\gamma_{\mathrm{cg}}(G)$ of G, in the S-game the corresponding invariant is denoted by $\gamma'_{\mathrm{cg}}(G)$.

This variant of the domination game was introduced by Borowiecki, Fiedorowicz, and Sidorowicz in [8]. In addition, as a tool for studying the connected domination game, they introduced a related game as follows. The *connected domination game with Chooser* has rules as the normal game, except that there is another player, Chooser, who can make zero, one, or more moves after any move of Dominator or Staller. A move of Chooser must also preserve the connectivity, but need not dominate any new vertices. Chooser has no specific goal, he can help either Dominator or Staller. The following result is the key for the applicability of this auxiliary game.

Lemma 5.10 (Chooser Lemma) [8, Lemma 1] *Consider the connected domination game with Chooser on a graph G. Suppose that in the game Chooser picks k vertices, and that both Dominator and Staller play optimally. Then at the end of the game the number of played vertices is at most $\gamma_{\mathrm{cg}}(G) + k$ and at least $\gamma_{\mathrm{cg}}(G) - k$.*

In some situations, Lemma 5.10 can be considered as a substitution for the Continuation Principle (Lemma 2.1) which does not hold for the connected domination game. Indeed, as a simple example consider the path on at least five vertices and assume that its set A of degree-2 vertices is declared to be dominated. Then the connected domination game played on $G|A$ can never be finished.

Let G be a graph. The smallest cardinality of a set of vertices of G that is both connected and dominating is the *connected domination number* $\gamma_{\mathrm{c}}(G)$ of G. The following result is parallel to Theorem 1.1, but the proof of its upper bound requires a careful consideration; see Bujtás, Dokyeesun, Iršič, and Klavžar [33, Theorem 2.1].

Theorem 5.11 [8, Theorem 1] *If G is a graph, then $\gamma_{\mathrm{c}}(G) \leq \gamma_{\mathrm{cg}}(G) \leq 2\gamma_{\mathrm{c}}(G) - 1$.*

Recall that k-trees are recursively defined as follows. The complete graph on k vertices is a k-tree, and if G is a k-tree, then the graph obtained by adding a new vertex adjacent to the vertices of a k-clique of G is also a k-tree. We have the following upper bound on the game connected domination number of 2-trees.

Theorem 5.12 [8, Theorem 4] *If G is a 2-tree of order $n \geq 4$, then*

$$\gamma_{\mathrm{cg}}(G) \leq \left\lceil \frac{2(n-4)}{3} \right\rceil + 1.$$

For the S-game, the following holds true.

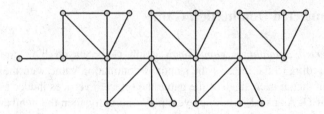

Fig. 5.2 The graph G_6.

Theorem 5.13 [8, Theorem 8] *If G is a graph, then $\gamma'_{cg}(G) \geq \gamma_{cg}(G) - 1$.*

Borowiecki et al. [8, Problem 1] also posed the problem to find the maximum k for which there exists a graph G satisfying $\gamma'_{cg}(G) = \gamma_{cg}(G) + k$. Iršič [79] showed that such k does not exist by constructing an infinite family of graphs G_n, $n \geq 2$, such that $\gamma_{cg}(G_n) = n$ [79, Lemma 3.3] and $\gamma'_{cg}(G_n) = 2n$ [79, Lemma 3.4]. We do not explicitly describe the graphs G_n here, but display the graph G_6 in Figure 5.2 from which the reader can guess the construction.

On the other hand, for a fixed graph we have the following bound.

Theorem 5.14 [79, Theorem 3.2] *If G is a graph, then $\gamma'_{cg}(G) \leq 2\gamma_{cg}(G)$.*

Iršič also considered the effect of predomination of a vertex on the game connected domination number. Moreover, in [79, Proposition 2.2] she observed that if G is a graph, then $\gamma_{cg}(G) \geq \mathrm{diam}(G) - 1$ and $\gamma'_{cg}(G) \geq \mathrm{diam}(G)$.

The connected domination game was also investigated on Cartesian products, for which we have the following general upper bound.

Theorem 5.15 [33, Theorem 2.2] *If G and H are connected graphs, then*

$$\gamma_{cg}(G \square H) \leq \min\{2\gamma_c(G)n(H), 2\gamma_c(H)n(G)\} - 1.$$

The bound of Theorem 5.15 is sharp. This in particular follows from [8, Theorems 6 and 7] where it is proved that if $2 \leq m \neq n \geq 4$, then $\gamma_{cg}(K_{1,n-1} \square K_m) = \min\{2n - 1, 2m - 1\}$, and if $m, n \geq 4$, then $\gamma_{cg}(P_n \square K_m) = 2n - 1$. Several additional exact results on $\gamma_{cg}(G \square H)$ are known; we point out the following two.

Theorem 5.16 [33, Theorem 3.1] *If $n \geq 3$ and $m \geq 1$, then $\gamma_{cg}(K_{1,n} \square P_m) = 2m - 1$.*

Theorem 5.17 [33, Theorem 5.1] *If G is a connected graph and $k \geq \min\{4\Delta(G) + \alpha(G), 2n(G) - 1\}$, then $\gamma_{cg}(K_k \square G) = 2n(G) - 1$.*

The following result might appear obvious, but its proof is non-trivial.

Theorem 5.18 [33, Theorem 5.2] *If $n \geq 1$, then $\gamma_{cg}(K_{n+1} \square G) \geq \gamma_{cg}(K_n \square G)$.*

Note that Theorem 5.18 does not generalize to general subgraphs. Indeed, if G is a graph with a large $\gamma_{\mathrm{cg}}(G)$ and H is the graph obtained from G by adding a dominating vertex, then $1 = \gamma_{\mathrm{cg}}(H) \ll \gamma_{\mathrm{cg}}(G)$.

The connected domination game was also studied on the lexicographic product of two graphs, for which the following holds for the S-game.

Theorem 5.19 [8, Theorem 4.6] *If G and H are graphs, then*

$$
\gamma_{\mathrm{cg}}'(G \circ H) = \begin{cases} \gamma_{\mathrm{cg}}'(G); & \gamma_{\mathrm{cg}}'(G) \geq 2, \\ 2; & \gamma_{\mathrm{cg}}'(G) = 1, n(G) \geq 2 \text{ and } \gamma_{\mathrm{cg}}'(H) \geq 2, \\ \gamma_{\mathrm{cg}}'(H); & n(G) = 1, \text{ or } \gamma_{\mathrm{cg}}'(G) = 1, n(G) \geq 2 \text{ and } \gamma_{\mathrm{cg}}'(H) = 1. \end{cases}
$$

A similar result holds also for the D-game [79, Theorem 4.4]. To state it explicitly one needs an auxiliary *Staller-first-skip connected domination game*, hence we skip the details here.

As the final thought on the connected domination game we add that in most situations (but not all!) this game seems to be easier than the usual domination game. Results like Theorems 5.16–5.19, for which no related result is known for the domination game, support this statement.

5.4 Z-Game, L-Game, and LL-Game

In Sections 4.5 and 4.6, two additional types of dominating sequences, complementing the closed and the open neighborhood sequences, were introduced, namely, the Z-sequences and the L-sequences. In the same way as the domination game is related to the closed neighborhood sequences and the total domination game is related to the open neighborhood sequences, one can also consider a game based on the condition of the Z-sequence and the L-sequence, respectively, by following essentially the same set of rules as in the (total) domination game. This idea was explored for the first time by Brešar, Bujtás, Gologranc, Klavžar, Košmrlj, Marc, Patkós, Tuza, and Vizer [11].

Each version of the game as introduced in [11] is played by Dominator and Staller on a graph G with no isolated vertices. The goals of Dominator and Staller are the same as in the standard domination game (Dominator's goal is to finish the game as soon as possible, while Staller wants just the opposite), and the terms D-game and S-game should also be clear. Now, when a *Z-domination game* is played on G, players are taking turns choosing a vertex v_i of G such that the condition $N(v_i) - \bigcup_{j=1}^{i-1} N[v_j] \neq \emptyset$ is true. When considering a game with respect to L-sequences, two games appear naturally. Recall that in the condition of L-sequences,

$$N[v_i] - \bigcup_{j=1}^{i-1} N(v_j) \neq \emptyset, \tag{5.2}$$

we additionally require that all chosen vertices are pairwise distinct, for otherwise the longest L-sequence would be infinite. However, when a game is played based on the same condition, we may not need to require that all chosen vertices are pairwise distinct, since Dominator will eventually prevent the repetition of vertices. Thus, in an *L-domination game*, the players must obey condition (5.2), and, in addition, $v_i \neq v_j$ if $i \neq j$, whereas in an *LL-domination game*, the players just need to obey condition (5.2), while repetitions of vertices are allowed.

If a D-game is played and both players play optimally, then the total number of moves played during the game is, respectively,

- the *game Z-domination number*, $\gamma_{Zg}(G)$;
- the *game L-domination number*, $\gamma_{Lg}(G)$; and
- the *game LL-domination number*, $\gamma_{LLg}(G)$, of G.

For the S-game, the total number of moves gives analogous graph invariants, denoted by $\gamma'_{Zg}(G)$, $\gamma'_{Lg}(G)$, and $\gamma'_{LLg}(G)$, respectively.

The Continuation Principle extends to all the mentioned invariants [11, Theorem 2.1].

Theorem 5.20 (Continuation Principles) *If G is a graph without isolated vertices and $B \subseteq A \subseteq V(G)$, then*

(i) $\gamma_{Zg}(G|A) \leq \gamma_{Zg}(G|B)$ and $\gamma'_{Zg}(G|A) \leq \gamma'_{Zg}(G|B)$;
(ii) $\gamma_{LLg}(G|A) \leq \gamma_{LLg}(G|B)$ and $\gamma'_{LLg}(G|A) \leq \gamma'_{LLg}(G|B)$; and
(iii) $\gamma_{Lg}(G|A) \leq \gamma_{Lg}(G|B)$ and $\gamma'_{Lg}(G|A) \leq \gamma'_{Lg}(G|B)$.

The following result shows that the difference between the length of the D-game and the S-game, when both players play optimally, is bounded by 1 for all three types of games. While the proofs for the Z-domination game and the LL-domination game use a standard application of the Continuation Principles, the proof for the L-domination game is more involved and also uses the imagination strategy.

Theorem 5.21 [11, Theorem 2.2] *If G is a graph without isolated vertices, then*

(i) $|\gamma_{Zg}(G) - \gamma'_{Zg}(G)| \leq 1$;
(ii) $|\gamma_{LLg}(G) - \gamma'_{LLg}(G)| \leq 1$; and
(iii) $|\gamma_{Lg}(G) - \gamma'_{Lg}(G)| \leq 1$.

The hierarchy of the five versions of the game domination numbers is presented in the next result of Brešar et al. [11]. (Note that γ_g and γ_{tg} are incomparable; recall Theorems 1.4 and 1.5.)

Theorem 5.22 [11, Theorem 3.1] *If G is a graph without isolated vertices, then*

$$\gamma_{Zg}(G) \leq \gamma_g(G) \leq \gamma_{Lg}(G) \leq \gamma_{LLg}(G) \quad \text{and} \quad \gamma_{Zg}(G) \leq \gamma_{tg}(G) \leq \gamma_{Lg}(G) \leq \gamma_{LLg}(G).$$

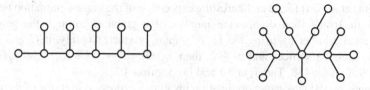

Fig. 5.3 Trees with pairwise distinct values of the game domination numbers.

All inequalities in Theorem 5.22 are sharp. Indeed, one example is the 5-cycle, where $\gamma_{Zg}(C_5) = \gamma_g(C_5) = \gamma_{tg}(C_5) = \gamma_{Lg}(C_5) = 3$, though $\gamma_{LLg}(C_5) = 5$. On the other hand, for the corona $K_n \odot \overline{K_n}$ of the complete graph K_n, $n \geq 2$, and its complement, we have $\gamma_g(K_n \odot \overline{K_n}) = \gamma_{Lg}(K_n \odot \overline{K_n}) = \gamma_{LLg}(K_n \odot \overline{K_n}) = 2n - 1$ (in this case, $\gamma_{Zg}(K_n \odot \overline{K_n}) = n$ and $\gamma_{tg}(K_n \odot \overline{K_n}) = n + 1$). In [11, Proposition 6.1] it was shown that for the Cartesian product of a connected nontrivial graph G with a star having k leaves, where $k \geq 2n(G)$, equality holds throughout the chain of inequalities in Theorem 5.22. In particular, $\gamma_{Zg}(G \square K_{1,k}) = \gamma_{LLg}(G \square K_{1,k}) = 2n(G) - 1$.

Examples showing that the values in the chain of inequalities in Theorem 5.22 can all be different were also presented in [11]. In fact, the smallest trees demonstrating this property were given. The left tree on 11 vertices in Figure 5.3 has the following values: $\gamma_{Zg} = 5$, $\gamma_g = 6$, $\gamma_{tg} = 7$, $\gamma_{Lg} = 8$, $\gamma_{LLg} = 9$, while the right tree on 14 vertices in the same figure has the same separability property, except that γ_g and γ_{tg} are reversed, and has the following values: $\gamma_{Zg} = 5$, $\gamma_{tg} = 6$, $\gamma_g = 7$, $\gamma_{Lg} = 8$, $\gamma_{LLg} = 9$.

Bujtás, Iršič, and Klavžar, [38] further explored the relations between the five game domination invariants. They presented another family of (Cartesian products of) graphs in which all five invariants coincide. On the other hand, they proved [38, Theorem 5.2] that $\gamma_{Zg}(G) + 1 \leq \gamma_{Lg}(G)$ if $\gamma_{Zg}(G)$ is an even number. In addition, they provided some sufficient conditions for a graph G to satisfy $\gamma_{Zg}(G) = \gamma_g(G)$ and $\gamma'_{Zg}(G) = \gamma'_g(G)$; see [38, Theorems 4.1 and 4.2].

A number of problems concerning relations between the five domination games remain open. For instance, it would be interesting to characterize the graphs G without isolated vertices for which $\gamma_{Zg}(G) = \gamma_{LLg}(G)$, respectively, $\gamma_{Zg}(G) = \gamma_{Lg}(G)$, holds [11, Problems 6.2 and 6.4]. The question whether $\gamma_{LLg}(G) \leq 2\gamma_{Zg}(G) + 1$ holds for an arbitrary graph G without isolated vertices was also posed in [11, Problem 6.5]. Concerning the relations between $\gamma_{Zg}(T)$ and $\gamma_{LLg}(T)$, Brešar et al. proposed the following conjecture involving the class of trees.

Conjecture 5.23 [11, Conjecture 6.3] If T is a nontrivial tree, then $\gamma_{Zg}(T) < \gamma_{LLg}(T)$.

Conjecture 5.23 was further strengthened by Bujtás et al.

Conjecture 5.24 [38, Conjecture 5.3] If T is a nontrivial tree, then $\gamma_{Zg}(T) < \gamma_{Lg}(T)$.

Bujtás et al. [38] found an interesting expression of the game domination number and of the game total domination number of a graph G through the game Z-domination number by using the lexicographic product. Notably, if G is a graph with no isolated vertices and $n \geq 2$, then $\gamma_{\mathrm{tg}}(G) = \gamma_{\mathrm{Zg}}(G \circ \overline{K_n})$, and $\gamma_{\mathrm{g}}(G) = \gamma_{\mathrm{Zg}}(G \circ K_n)$; see [38, Theorem 3.1 and Proposition 3.2].

Bounds on the new game domination invariants expressed in terms of the (total) domination number of a graph were also proved.

Proposition 5.25 [11, Proposition 4.1] *If G is a graph without isolated vertices, then*

(i) $\gamma(G) \leq \gamma_{\mathrm{Zg}}(G) \leq 2\gamma(G) - 1$;
(ii) $\gamma_t(G) \leq \gamma_{\mathrm{Lg}}(G) \leq 2\gamma_t(G) - 1$; *and*
(iii) $\gamma_t(G) + 1 \leq \gamma_{\mathrm{LLg}}(G) \leq 2\gamma_t(G) - 1$.

For the largest among the domination game invariants, the order of a graph need not be an upper bound:

Theorem 5.26 [11, Theorem 4.4] *If G is a graph without isolated vertices, then $\gamma_{\mathrm{LLg}}(G) \leq n(G) + 1$. Moreover, equality holds if and only if all components of G are graphs K_2.*

Nevertheless, if G is a graph with no isolated vertices such that not all of its components are K_2, Theorem 5.26 implies $\gamma_{\mathrm{LLg}}(G) \leq n(G)$. It would be interesting to characterize the graphs that attain the equality. This problem is also open and perhaps easiest in the class of trees [11, Problem 6.6].

On the other hand, obtaining a sharp upper bound for the game L-domination number in terms of a fraction of the order of a graph provides another challenge. The $\frac{6}{7}$-**Conjecture** was posed in [11, Conjecture 6.7].

Conjecture 5.27 If G is a graph without isolated vertices, then $\gamma_{\mathrm{Lg}}(G) \leq \frac{6}{7}n(G)$.

The conjecture has been verified by computer for all trees on up to 18 vertices, and an infinite family of graphs G attaining $\gamma_{\mathrm{Lg}}(G) = \frac{6}{7}n(G)$ was presented in [11, Proposition 6.8].

The three new game invariants were studied also on paths. For the L-game it was proved [11, Theorem 5.4] that for every positive integer n there exists a constant c_n such that $\gamma_{\mathrm{Lg}}(P_n) = \frac{2n}{3} + c_n$ holds with $|c_n| \leq 1$. For the LL-game, $\gamma_{\mathrm{LLg}}(P_n) = \frac{4n}{5} + c_n$ for some small bounded constants c_n, as proved in [11, Theorem 5.5]. A similar result for the Z-game on paths in [11] was improved to the exact value in [38, Corollary 4.3].

5.5 Transversal Game in Hypergraphs

In 2016, Bujtás, Henning, and Tuza [34] introduced and first studied the transversal game in hypergraphs. A subset T of vertices in a hypergraph H is a *transversal* (also called *hitting set* or *blocking set* in many papers) if T has a nonempty intersection with every edge of H. The minimum cardinality of a transversal set in H is the *transversal number*, $\tau(H)$, of H. A vertex *hits* or *covers* an edge if it belongs to that edge. As defined in [34], the transversal game played on a hypergraph H consists of two players, *Edge-hitter* and *Staller*, who take turns choosing a vertex from H. Each vertex chosen must hit at least one edge not hit by the vertices previously chosen. The game ends when the set of vertices chosen becomes a transversal in H. Edge-hitter wishes to end the game with a minimum number of vertices chosen, and Staller wishes to end the game with as many vertices chosen as possible. The *game transversal number* (resp. *Staller-start game transversal number*), $\tau_g(H)$ (resp. $\tau_g'(H)$), of H is the number of vertices chosen when Edge-hitter (resp. Staller) starts the game and both players play optimally according to their goals.

Since the transversal game played in a hypergraph H ends when the set of vertices chosen becomes a transversal in H, we have $\tau(H) \leq \tau_g(H)$ and $\tau(H) \leq \tau_g'(H)$. If Edge-hitter fixes a minimum transversal set, T, in H and adopts the strategy in each of his turns to play a vertex from T if possible, then he guarantees that the game ends in no more than $2\tau(H) - 1$ moves in the Edge-hitter-start transversal game and in no more than $2\tau(H)$ moves in the Staller-start transversal game. We state this fact formally as follows.

Observation 5.28 [34, Observation 1] *For every hypergraph H, the following holds.*

(a) $\tau(H) \leq \tau_g(H) \leq 2\tau(H) - 1$,
(b) $\tau(H) \leq \tau_g'(H) \leq 2\tau(H)$.

As remarked in [34], the equalities $\tau(H) = \tau_g(H) = \tau_g'(H)$ hold if H is the disjoint union of complete k-uniform hypergraphs. Further, $\tau_g(H) = 2\tau(H) - 1$ and $\tau_g'(H) = 2\tau(H)$ are valid if $\tau(H) = 1$ and H contains at least two different edges. The following infinite family of hypergraphs H with $\tau_g(H) = 2\tau(H) - 1$ and $\tau_g'(H) = 2\tau(H)$ was presented by Bujtás, Henning, and Tuza in [35]. A k-*corona* of a hypergraph H is a hypergraph obtained by attaching k edges (each of size at least 2) to each vertex of H, where the edges attached to a vertex $v \in V(H)$ contain only degree-1 vertices apart from v. We are now in a position to present the following result, which shows that the lower and upper bounds given in Observation 5.28 cannot be improved even when $\tau(H)$ is large.

Proposition 5.29 [35, Proposition 1] *For every positive integer k and for every hypergraph H of order at most $2^{k-1} - 1$, every k-corona H^k of H satisfies $\tau_g(H^k) = 2\tau(H^k) - 1$ and $\tau_g'(H^k) = 2\tau(H^k)$.*

The Continuation Principle also holds for the transversal game. This fundamental lemma expresses the monotonicity of τ_g and τ_g' with respect to subhypergraphs; see [35, Lemma 3]. By $H|A$ we denote a hypergraph together with a declaration that the edges from a subset A are already covered.

Lemma 5.30 (Transversal Continuation Principle) *Let H be a hypergraph and let $A, B \subseteq E(H)$. If $B \subseteq A$, then $\tau_g(H|A) \leq \tau_g(H|B)$ and $\tau_g'(H|A) \leq \tau_g'(H|B)$.*

The Transversal Continuation Principle can be stated in the following equivalent form [35, Lemma 3′].

Lemma 5.31 (Transversal Continuation Principle) *Let H_1 and H_2 be two hypergraphs on the same vertex set with edge sets E_1 and E_2, respectively. If $E_1 \subseteq E_2$, then $\tau_g(H_1) \leq \tau_g(H_2)$ and $\tau_g'(H_1) \leq \tau_g'(H_2)$.*

During the course of the transversal game, the edges which are already covered and the vertices which are not incident with any uncovered edges do not influence the continuation of the game. Hence, these are deleted to obtain the residual hypergraph. As a consequence of the Transversal Continuation Principle, if a vertex v hits/covers all the edges hit by a vertex u in the residual hypergraph H where $d_H(v) > d_H(u)$, then we may suppose Edge-hitter never plays u and Staller never plays v.

As another consequence of the Transversal Continuation Principle, the number of moves in the Edge-hitter-start transversal game and the Staller-start transversal game when played optimally can differ by at most one. We state this formally as follows.

Theorem 5.32 [35, Theorem 4] *If H is a hypergraph, then $|\tau_g(H) - \tau_g'(H)| \leq 1$.*

A tight upper bound on the game transversal number of a hypergraph in terms of its order and size was given in [34]. In order to state the result, we define M_1 to be the hypergraph with vertex set $V(M_1) = \{x_1, x_2, x_3, y_1, y_2, y_3\}$ and edge set

$$E(M_1) = \{\{x_1, x_2, x_3\}, \{y_1, y_2, y_3\}, \{x_1, y_1\}, \{x_2, y_2\}, \{x_3, y_3\}\}.$$

For $k \geq 1$, let M_k consist of k disjoint copies of M_1, and let $\mathcal{H} = \{M_k : k \geq 1\}$. The hypergraph M_2 is illustrated in Figure 5.4, albeit without the vertex labels.

We are now in a position to state the following upper bound on the game transversal number of a hypergraph. We shall use the notation $n_H = |V(H)|$ and $m_H = |E(H)|$ to denote the order and size of a hypergraph H, respectively.

Theorem 5.33 [34, Theorem 1] *If H is a hypergraph with all edges of size at least 2, and $H \not\cong C_4$, then $\tau_g(H) \leq \frac{4}{11}(n_H + m_H)$, with equality if and only if $H \in \mathcal{H}$.*

Given a graph G denote by H_G the open neighborhood hypergraph (ONH for short) of G. We note that if G has order n, then H_G has $n_H = n$ vertices and

Fig. 5.4 The hypergraph, M_2, in the family $\mathcal{H}$.

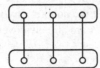 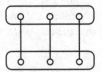

$m_H = n$ edges. The transversal number of the ONH of a graph is precisely the total domination number of the graph; that is, for a graph G, we have $\gamma_t(G) = \tau(H_G)$.

We next describe an interplay between the game total domination number and the game transversal number as first observed in [34]. A sequence of moves in the total domination game is legal if and only if the sequence of moves is legal in the transversal game. Thus, there is a one-to-one correspondence between the sequences of legal moves in the total domination game and the sequences of legal moves in the transversal game, implying the following observation.

Observation 5.34 [34, Observation 1] *If G is a graph with no isolated vertex and H_G is the ONH of G, then $\gamma_{tg}(G) = \tau_g(H_G)$.*

Let G be a graph of order n with $\delta(G) \geq 2$ and let $H = H_G$ be the ONH of G. Then, each edge of H has size at least 2. Since C_4 and any $H \in \mathcal{H}$ is not the ONH of any graph, by Theorem 5.33 and Observation 5.34 we obtain

$$\gamma_{tg}(G) = \tau_g(H) < \frac{4}{11}(n_H + m_H) = \frac{4}{11}(n + n) = \frac{8}{11}n.$$

We state this result, which demonstrates another example of the phenomenon that a general problem may turn out to be easier to handle than its particular case, formally as follows.

Theorem 5.35 [34, Corollary 1] *If G is a graph with $\delta(G) \geq 2$, then $\gamma_{tg}(G) \leq \frac{8}{11}n(G)$.*

As a special case of more general results due to Tuza [108] and Chvátal and McDiarmid [49], if H is a 2-uniform hypergraph, then $\tau(H) \leq \frac{1}{3}(n_H + m_H)$. This bound is almost true for the game transversal number as shown in [34].

Theorem 5.36 [34, Theorem 2 and Corollary 3] *If H is a 2-uniform hypergraph, then $\tau_g(H) \leq \frac{1}{3}(n_H + m_H + 1)$ and $\tau_g'(H) \leq \frac{1}{3}(n_H + m_H + 2)$.*

The game transversal number of a cycle and path is determined as follows.

Proposition 5.37 [34, Propositions 3 and 4] *The following holds.*

(a) *If $n \geq 3$, then $\tau_g(C_n) = \left\lfloor \frac{2n+1}{3} \right\rfloor$ and $\tau_g'(C_n) = \left\lfloor \frac{2n}{3} \right\rfloor$.*

(b) *If $n \geq 2$, then $\tau_g'(P_n) = \left\lfloor \frac{2n}{3} \right\rfloor$ and $\tau_g(P_n) = \left\lfloor \frac{2n-1}{3} \right\rfloor$.*

By Proposition 5.37, if $H = C_n$ where $n \equiv 1 \pmod 3$, then $\tau_g(H) = \frac{1}{3}(2n + 1) = \frac{1}{3}(n_H + m_H + 1)$. Thus, the first bound of Theorem 5.36 is achieved by cycles of length congruent to 1 modulo 3. Similarly, the second bound is achieved by cycles of length congruent to 0 modulo 3.

While the tight bounds $\tau(H) \leq \frac{1}{3}(n_H + m_H)$ and $\tau_g(H) \leq \frac{1}{3}(n_H + m_H + 1)$ on graphs H (or, equivalently, 2-uniform hypergraphs) are quite similar, the behaviors of τ and τ_g become substantially different if we restrict our attention to connected graphs. Although the bound on τ_g remains the same, demonstrated by Theorem 5.36, a much stronger inequality $\tau(H) \leq \frac{2}{7}(n_H + m_H + 1)$ is valid (which is again tight), as proved in [57]. In other words, in terms of $n_H + m_H$, on connected graphs H the best possible asymptotic coefficient for τ is $\frac{2}{7}$, whereas that for τ_g is $\frac{1}{3}$.

The main results in [35] show that the upper bound of Theorem 5.33 can be improved for 3-uniform and 4-uniform hypergraphs as follows.

Theorem 5.38 [35, Theorem 5 and Corollary 2] *If H is a 3-uniform hypergraph, then the following holds:*

(a) $\tau_g(H) \leq \frac{5}{16}(n_H + m_H)$,
(b) $\tau'_g(H) \leq \frac{1}{16}(5n_H + 5m_H + 6)$.

Theorem 5.39 [35, Theorem 6 and Corollary 4] *If H is a 4-uniform hypergraph, then the following holds.*

(a) $\tau_g(H) \leq \frac{71}{252}(n_H + m_H)$,
(b) $\tau'_g(H) \leq \frac{1}{252}(71n_H + 71m_H + 110)$.

It remains an open problem to determine whether the bounds in Theorem 5.38 and Theorem 5.39 are sharp. We close this section with the following problem.

Problem 5.40 [35, Problem 3] If $k \geq 5$, find the minimum c_k for which every k-uniform hypergraph H satisfies $\tau_g(H) \leq c_k(n_H + m_H)$.

5.6   (Total) Domination Game in Hypergraphs

In Section 5.5, we discussed the transversal game in hypergraphs. In this section, we turn our attention to the domination game in hypergraphs, which was introduced and first studied in 2019 by Bujtás, Patkós, Tuza, and Vizer [44].

Before we formally define the domination game in hypergraphs, let us recall some standard hypergraph notation. Two vertices x and y in a hypergraph H are *adjacent* if there is an edge e of H such that $\{x, y\} \subseteq e$. The *neighborhood* of a vertex v in H, denoted $N_H(v)$, is the set of all vertices different from v that are adjacent to v, while the *closed neighborhood* of v is the set $N_H[v] = N_H(v) \cup \{v\}$. We call a vertex in $N_H(v)$ a *neighbor* of v in H. For a subset $S \subseteq V(H)$, the *neighborhood* of S is the set $N_H(S) = \cup_{x \in S} N_H(x)$, while the *closed neighborhood* of S is the set $N_H[S] = N_H(S) \cup S$.

A vertex *dominates* itself and all its neighbors. A *dominating set* in a hypergraph H is a subset of vertices $D \subseteq V(H)$ such that every vertex is dominated by at least one vertex in D, that is, every vertex $v \in V(H) - D$ is adjacent to a vertex in the set D. Equivalently, $D \subseteq V(H)$ is a dominating set if $N[D] = V(H)$. The *domination number* $\gamma(H)$ is the minimum cardinality of a dominating set in H.

The domination game in hypergraphs is analogous to that on graphs, and is played on a hypergraph H by Dominator and Staller, who take turns choosing a vertex from H. Whenever a vertex is chosen it must dominate at least one vertex not dominated by the vertices previously chosen. The game ends when no move is possible. When this occurs, the set of selected vertices is a dominating set of G. We denote by $v_1, \ldots, v_d$ the sequence of vertices selected by the players in the game, and let $D_i = \{v_j : j \in [i]\}$ for all $i \in [d]$. Thus, the sequence $v_1, \ldots, v_d$ of played vertices satisfies $N[v_i] - N[D_{i-1}] \neq \emptyset$ for all $2 \leq i \leq d$. Further, the set D_d is a dominating set of H. We call the sequence $v_1, \ldots, v_d$ a *legal sequence* in the domination game played in H.

Dominator wishes that the game is ended in as few moves as possible, and Staller wishes to maximize the total number of moves. We adopt here exactly our notation for the domination game in graphs. When both players play optimally and Dominator (resp. Staller) starts the game, the uniquely defined length of the game (that is the number of moves altogether played in the game or equivalently the number of chosen vertices) is the *game domination number*, $\gamma_g(H)$ (resp. the *Staller-start game domination number*, $\gamma_g'(H)$) of H.

Analogously as in the domination game in graphs, given a hypergraph H and a subset $S \subseteq V(H)$, we denote by $H|S$ the *partially dominated hypergraph*, in which vertices contained in S are declared to have been already dominated. That is, D is a dominating set of $H|S$ if $N_H[D] \cup S = V(H)$. Adopting our notation for graphs, we write $\gamma_g(H|S)$ (resp. $\gamma_g'(H|S)$) to denote the number of moves remaining in the domination game on $H|S$ under optimal play when Dominator (resp. Staller) has the next move. Two games on hypergraphs H_1 and H_2 are called *equivalent* if $V(H_1) = V(H_2)$ and any sequence $v_1, \ldots, v_d$ is a legal sequence in the domination game played on H_1 if and only if it is a legal sequence in the domination game played on H_2.

For a hypergraph H, the *closed neighborhood hypergraph* $\mathrm{CNH}(H)$ of H is the hypergraph defined on the same set $V(H)$ of vertices, whose edges are the closed neighborhoods of vertices of H, that is, $E(\mathrm{CNH}(H)) = \{N_H[v] : v \in V(H)\}$. We note that the closed neighborhood hypergraph $\mathrm{CNH}(H)$ of a hypergraph H is a multi-hypergraph, since different vertices of H may have the same neighborhood in H.

The *2-section graph* $[H]_2$ of a hypergraph H is the graph whose vertex set is $V(H)$ and where two vertices u and v are adjacent in $[H]_2$ if and only if they are adjacent in H (that is, u and v belong to a common edge of H). By definition, we note that for every vertex $v \in V(H)$, the set of neighbors of v in the hypergraph H is precisely the set of neighbors of v in the 2-section graph $[H]_2$ of H.

The domination game in hypergraphs is related to the domination game in graphs and the transversal game in hypergraphs as follows.

Proposition 5.41 [44, Proposition 2.1] *If H is a hypergraph and $S \subseteq V(H)$, then the following hold.*

(a) *The domination game on H is equivalent to the domination game on $[H]_2$ and to the transversal game on $\mathrm{CNH}(H)$.*

(b) $\gamma_g(H) = \gamma_g([H]_2) = \tau_g(\mathrm{CNH}(H))$.

(c) $\gamma_g(H|S) = \gamma_g([H]_2|S)$.

We state next the Continuation Principle for the domination game in hypergraphs [44, Corollary 2.2].

Lemma 5.42 (Hypergraph Domination Continuation Principle) *If H is a hypergraph and $A, B \subseteq V(H)$ with $B \subseteq A$, then $\gamma_g(G|A) \leq \gamma_g(G|B)$ and $\gamma_g'(G|A) \leq \gamma_g'(G|B)$.*

As a consequence of Theorems 2.20 and 2.24, and Proposition 5.41, we have the following bounds on the game domination number in hypergraphs.

Corollary 5.43 *If H is a hypergraph of order n without isolated vertices, then the following hold.*

(a) *If every edge of H has size at least 2, then $\gamma_g(H) \leq \frac{5}{8}n$.*
(b) *If every edge of H has size at least 3, then $\gamma_g(H) \leq \frac{3}{5}n$.*

The bound in Corollary 5.43(b) was improved for 3-uniform hypergraphs as follows.

Theorem 5.44 [44, Theorems 3.4 and 5.3] *If H is an isolate-free 3-uniform hypergraph of order n, then $\gamma_g(H) \leq \frac{5}{9}n$ and $\gamma_g'(H) \leq \frac{5}{9}n$.*

The proof of Theorem 5.44 is very beautiful, and uses the *Bujtás discharging method* (illustrated in the proof Theorem 2.10) that assigns colors and weights to the vertices. The proof studies two different phases of the game and the possible structures of the residual hypergraph in each phase.

As a consequence of Theorem 5.44, we have the following results.

Corollary 5.45 [44, Theorem 6.1] *If H is a hypergraph of order n without isolated vertices and such that every edge of H has size at least 3, then $\gamma_g(H) \leq \frac{5}{9}n$.*

Corollary 5.46 [44, Theorems 3.4 and 5.3] *If G is an isolate-free graph of order n and each edge of G belongs to a triangle, then $\gamma_g(G) \leq \frac{5}{9}n$ and $\gamma_g'(G) \leq \frac{5}{9}n$.*

It remains, however, an open problem to determine a best possible upper bound on the game domination number of a 3-uniform hypergraph without isolated vertices in terms of its order.

Problem 5.47 Find the smallest constant C for which every 3-uniform hypergraph H of order n without isolated vertices satisfies $\gamma_g(H) \leq C \cdot n$.

Bujtás et al. [44] showed that $C \geq \frac{4}{9}$. To see this, consider the D-game played in the hypergraph H consisting of all horizontal and vertical lines except one of the 3×3 grid. Formally, let $V(H) = [9]$ and $E(H) = \{\{1, 2, 3\}, \{4, 5, 6\}, \{1, 4, 7\}, \{2, 5, 8\}, \{3, 6, 9\}\}$. (See also Figure 5.5.) Let H have order n, and so $n = 9$. Because of symmetry, we may assume that the first move of Dominator is either the vertex 1 or the vertex 7 (only the degree of the vertex picked matters). By the Hypergraph Domination Continuation Principle, we may further assume that Dominator chooses the vertex 1 on his first move. Staller responds by choosing the vertex 4, which dominates the vertices 5 and 6. After these two

Fig. 5.5 A 3-uniform hypergraph H of order n with $\gamma_g(H) = \gamma_g'(H) = \frac{4}{9}n$.

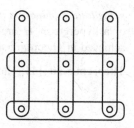

moves are played, there are exactly two vertices that are not yet dominated, namely vertices 8 and 9. However, these two vertices are non-adjacent, and therefore at least two further moves are required to dominate them. This shows that $\gamma_g(H) \geq 4$. On the other hand, if Dominator starts with the vertex 1 as before, then every vertex chosen by Staller on her first turn will dominate at least two new vertices. After these first two moves, at least seven vertices are therefore dominated, and hence at most two vertices are not yet dominated, implying that at most two additional moves are required to complete the game. This shows that $\gamma_g(H) \leq 4$. Consequently, $\gamma_g(H) = 4 = \frac{4}{9}n$, implying that $C \geq \frac{4}{9}$. We state this formally as follows.

Proposition 5.48 [44, Proposition 3.5] *There exists a 3-uniform hypergraph H of order n satisfying* $\gamma_g(H) = \gamma_g'(H) = \frac{4}{9}n$.

We next present an asymptotically tight upper bound on the ratio of the game domination number of a k-uniform hypergraph over its order ($o_k(1)$ being a function of k that converges to 0 when k tends to infinity).

Theorem 5.49 [44, Theorem 3.1] *If $k \geq 3$ and H is a k-uniform hypergraph, then*

$$\frac{\gamma_g(H)}{|V(H)|} \leq (2 + o_k(1))\frac{\log k}{k}.$$

Moreover, there exists a sequence $(H_k)_{k \in \mathbb{N}}$ of k-uniform hypergraphs such that

$$\frac{\gamma_g(H_k)}{|V(H_k)|} \geq (2 - o_k(1))\frac{\log k}{k}.$$

A *total dominating set* in a hypergraph H is a subset of vertices $D \subseteq V(H)$ such that every vertex in H is adjacent with a vertex in D. Equivalently, a set D is a total dominating set in H if D is a dominating set in H with the additional property that for every vertex $v \in D$ there exists an edge $e \in E(H)$ for which $v \in e$ and $e \cap (D - \{v\}) \neq \emptyset$. The *total domination number* $\gamma_t(H)$ is the minimum cardinality of a total dominating set in H.

The total version of the domination game in hypergraphs is defined analogously to the total version of the domination game in graphs which we studied in Chapter 3. If one uses open neighborhoods instead of closed ones in the definition of the game domination number, then we have the total domination game in hypergraphs. The *game total domination number* and the *Staller-start game total domination number* of a hypergraph H are denoted by $\gamma_{\mathrm{tg}}(H)$ and $\gamma'_{\mathrm{tg}}(H)$, respectively. The result achieved in Theorem 5.49 also holds for the game total domination number.

Theorem 5.50 [44, Theorem 3.2] *If $k \geq 3$ and H is a k-uniform hypergraph, then*

$$\frac{\gamma_{\mathrm{tg}}(H)}{|V(H)|} \leq (2 + o_k(1))\frac{\log k}{k}.$$

Moreover, there exists a sequence $(H_k)_{k \in \mathbb{N}}$ of k-uniform hypergraphs such that

$$\frac{\gamma_{\mathrm{tg}}(H_k)}{|V(H_k)|} \geq (2 - o_k(1))\frac{\log k}{k}.$$

5.7 Additional Related Domination Games

Several additional related domination games are studied in the literature. In this section, we briefly discuss them.

5.7.1 Maker-Breaker Domination Games

The Maker-Breaker game was introduced in 1973 by Erdős and Selfridge [56]. The game is played on a hypergraph by Maker and Breaker who take turns, and at each turn the current player selects a new vertex. Maker wins if at some point of the game he has selected all vertices from one of the edges, while Breaker wins if she can keep him from doing it.

Duchêne, Gledel, Parreau, and Renault [55] introduced the *Maker-Breaker domination game* as follows. The game is played on a graph G with two players which were named, to be consistent with the usual domination game, Dominator and Staller. The players alternatively select a vertex of G that was not yet chosen in the course of the game. Dominator wins if at some point, the vertices he has chosen form a dominating set. Staller wins if Dominator cannot form a dominating set.

Note that the Maker-Breaker domination game is an instance of the Maker-Breaker game. Indeed, if G is a graph and $\mathcal{H}_G$ a hypergraph with the same set of vertices as G, and in which the edges are the dominating sets of G, then Dominator wins the Maker-Breaker domination game on G if and only if Maker wins the Maker-Breaker game on $\mathcal{H}_G$.

In [55, Theorem 6], Duchêne et al. proved that deciding the outcome of a Maker-Breaker domination game is PSPACE-complete on bipartite graphs. On the positive side they proved that deciding the outcome of the Maker-Breaker domination game on cographs as well as on trees can be done in polynomial time.

The *Maker-Breaker domination number* $\gamma_{MB}(G)$ of a graph G is the minimum number of moves of Dominator to win the Maker-Breaker domination D-game provided that he has a winning strategy. Otherwise set $\gamma_{MB}(G) = \infty$. Similarly, $\gamma'_{MB}(G)$ denotes the minimum number of moves of Dominator in the Maker-Breaker domination S-game. These invariants were introduced by Gledel, Iršič, and Klavžar [61], where it was proved (see [61, Theorem 3.1]) that if G is a graph, then $\gamma(G) \leq \gamma_{MB}(G) \leq \gamma'_{MB}(G)$. Moreover, for any integers r, s, t, where $2 \leq r \leq s \leq t$, there exists a graph G such that $\gamma(G) = r$, $\gamma_{MB}(G) = s$, and $\gamma'_{MB}(G) = t$. This result indicated that the Maker-Breaker domination game is intrinsically different from the domination game, cf. Theorem 2.2. Among other results, Gledel, Iršič, and Klavžar [61, Theorem 4.5] determined $\gamma_{MB}(T)$ and $\gamma'_{MB}(T)$ for a tree T.

Gledel, Henning, Iršič, and Klavžar [60] also studied a Maker-Breaker total domination game. It is defined much the same as the Maker-Breaker domination game, except that Dominator wins on G if he can select a total dominating set of G, and Staller wins if she can select all the vertices from the open neighborhood of some vertex. In [60] the Maker-Breaker total domination game was studied mostly from the outcome point of view and from the complexity point of view. For instance, deciding the outcome of the Maker-Breaker total domination game is PSPACE-complete on split graphs [60, Theorem 6.1].

5.7.2 Enclaveless Game

If S is a set of vertices in a graph G, then a vertex $v \in S$ is an *enclave* of S if it and all of its neighbors are also in S; that is, if $N[v] \subseteq S$. A set S is *enclaveless* if it does not contain any enclaves. Note that a set S is a dominating set of a graph G if the set $V(G) - S$ is enclaveless. The *enclaveless number* of G, denoted $\Psi(G)$, is the maximum cardinality of an enclaveless set in G, and the *lower enclaveless number* of G, denoted by $\psi(G)$, is the minimum cardinality of a maximal enclaveless set. The domination and enclaveless numbers of a graph G are related by the following equations.

Observation 5.51 *If G is a graph, then $\gamma(G) + \Psi(G) = n(G) = \Gamma(G) + \psi(G)$.*

In 2001 Phillips and Slater [100, 101] introduced and first studied the *competition-enclaveless game*. The game is played by two players, Maximizer and Minimizer, on some graph G. They take turns in constructing a maximal enclaveless set S of G. On each turn a player chooses a vertex v that is not in the set S of the vertices already chosen and such that $S \cup \{v\}$ does not contain an enclave, until there is no such vertex. The goal of Maximizer is to make the final set S as large as possible and for Minimizer to make the final set S as small as possible.

The *competition-enclaveless game number*, or simply the *enclaveless game number*, $\Psi_g^+(G)$ of G is the number of vertices chosen when Maximizer starts the game and both players play an optimal strategy according to the rules. The *Minimizer-start competition-enclaveless game number*, or simply the *Minimizer-*

start enclaveless game number, $\Psi_g^-(G)$, of G is the number of vertices chosen when Minimizer starts the game and both players play an optimal strategy according to the rules.

Although the domination and enclaveless numbers of a graph G are related by the equation $\gamma(G) + \Psi(G) = n(G)$ (see Observation 5.51), as remarked in [67] the competition-enclaveless game is very different from the domination game. For $k \geq 3$, if G is a tree with exactly two non-leaf vertices both of which have k leaf neighbors, that is, if G is a double star $S(k,k)$, then $\gamma_g(G) = 3$ and $\Psi_g^+(G) = k+1$, whence $\gamma_g(G) + \Psi_g^+(G) = \frac{1}{2}n + 3$. However, if G is a path P_n on $n \geq 2$ vertices, then it follows from results due to Košmrlj [93] and Phillips and Slater [101] that $\gamma_g(G) + \Psi_g^+(G) \approx n + \frac{1}{10}n$. These two simple examples show that the sum $\gamma_g(G) + \Psi_g^+(G)$ on the class of graphs of a fixed order can differ greatly even when restricted to trees.

As remarked in [67], the most significant difference between the domination game defined in Chapter 2 and the competition-enclaveless game introduced by Phillips and Slater [100, 101] is that the Continuation Principle (see Lemma 2.1) holds for the domination game but does not hold for the competition-enclaveless game. In particular, this implies that for the domination game the numbers $\gamma_g(G)$ and $\gamma_g'(G)$ can differ by at most 1, but for the competition-enclaveless game the numbers $\Psi_g^+(G)$ and $\Psi_g^-(G)$ can vary greatly. For example, if $n \geq 1$ and G is a star $K_{1,n}$, then $\Psi_g^+(G) = n$ while $\Psi_g^-(G) = 1$.

As remarked by Henning and Rall in [74], another significant difference between the domination game and the competition-enclaveless game is that upon completion of the domination game, the set of played vertices is a dominating set although not necessarily a minimal dominating set, while upon completion of the competition-enclaveless game, the set of played vertices is always a maximal enclaveless set. Thus, the enclaveless game numbers of a graph G are always squeezed between the lower enclaveless number $\psi(G)$ of G and the enclaveless number $\Psi(G)$ of G. We state this formally as follows.

Observation 5.52 [74, Observation 2] *If G is a graph of order n, then*

$$\psi(G) \leq \Psi_g^-(G) \leq \Psi(G) \quad and \quad \psi(G) \leq \Psi_g^+(G) \leq \Psi(G).$$

A graph G is *well-dominated* if all the minimal dominating sets of G have the same cardinality. As a consequence of Observation 5.52, we have the following connection between the enclaveless game and the class of well-dominated graphs.

Observation 5.53 [74, Observation 3] *If G is a well-dominated graph of order n, then $\Psi_g^-(G) = \Psi_g^+(G) = n - \gamma(G)$.*

The following bounds on the (Maximizer-start) enclaveless game number and the Minimizer-start enclaveless game number are given in [74].

Theorem 5.54 [74] *If G is an isolate-free graph of order n with maximum degree $\Delta(G) = \Delta$, then*

$$\left(\frac{1}{\Delta+1}\right)n \le \Psi_g^-(G) \le \left(\frac{\Delta}{\Delta+1}\right)n \quad and \quad \left(\frac{1}{\Delta+1}\right)n \le \Psi_g^+(G) \le \left(\frac{\Delta}{\Delta+1}\right)n.$$

The lower bound in Theorem 5.54 on $\Psi_g^-(G)$ is achieved, for example, by taking $G = K_{1,\Delta}$ for any given $\Delta \ge 1$ in which case $\Psi_g^-(G) = 1 = (\frac{1}{\Delta+1})n$ where $n = n(G) = \Delta + 1$. As shown in [74], the upper bounds in Theorem 5.54 are realized for infinitely many connected graphs.

The following $\frac{1}{2}$-**Enclaveless Game Conjecture** was first posed as a question by Slater in 2015, and subsequently as a conjecture by Henning [67].

Conjecture 5.55 [67, Conjecture 12] If G is a connected graph of order $n \ge 2$, then $\Psi_g^+(G) \ge \frac{1}{2}n$.

Henning and Rall [74] posed the following conjecture for the Minimizer-start enclaveless game.

Conjecture 5.56 [74, Conjecture 2] If G is a graph of order n with $\delta(G) \ge 2$, then $\Psi_g^-(G) \ge \frac{1}{2}n$.

The authors in [74] show that the $\frac{1}{2}$-Enclaveless Game Conjecture holds for regular graphs and claw-free graphs. Further, they show that Conjecture 5.56 for the Minimizer-start enclaveless game holds for regular graphs, and for connected claw-free graphs even if we relax the minimum degree 2 condition and replace it with the requirement that the graph is isolate-free and different from the path P_3.

5.7.3 Independent Domination Game

In 2001 Phillips and Slater [100, 101] introduced the *competition-independence game*. Adopting the notation coined by Goddard and Henning [62], the game is played by two players, Diminisher and Sweller, on some graph G. They take turns in constructing a maximal independent set M of G. That is, on each turn a player chooses a vertex that is not adjacent to any of the vertices already chosen until there is no such vertex. Upon completion of the game, the resulting set of chosen vertices is a dominating set of G (since it forms a maximal independent set in G). The goal of Diminisher is to make the final set M as small as possible and for Sweller to make the final set M as large as possible.

For a graph G, let $I_d(G)$ denote the size of the maximal independent set in the competition-independence game if Diminisher moves first and both players play optimally, and let $I_s(G)$ denote the size of the maximal independent set if Sweller moves first and both players play optimally. These numbers are called the *competition-independence numbers*.

As with the competition-enclaveless game, the Continuation Principle does not hold for the competition-independence game. Indeed the competition-independence numbers can vary greatly on who goes first. For example, if the graph G is a star and

Diminisher goes first, then he will choose the center vertex and the game is over. In contrast, if Sweller goes first, then she chooses a leaf. In this case, the resulting set M consists of all but one vertex, namely all the leaves. That is, $I_d(K_{1,k}) = 1$ while $I_s(K_{1,k}) = k$ for $k \geq 1$.

Phillips and Slater [101] determined the competition-independence numbers of paths and cycles. If $n \geq 1$, then $I_d(P_n) = \lfloor \frac{3n+4}{7} \rfloor$ and $I_s(P_n) = \lfloor \frac{3n+5}{7} \rfloor$, while if $n \geq 3$, then $I_d(C_n) = \lfloor \frac{3n+3}{7} \rfloor$ and $I_s(C_n) = \lfloor \frac{3n+2}{7} \rfloor$.

The competition-independence game is very non-trivial even when played on trees with maximum degree at most 3. For the case that Diminisher moves first, a computer search shows that $I_d(G) \leq \lfloor n/2 \rfloor$ for all trees of maximum degree 3 up through order at least $n = 15$. However, Goddard and Henning [62] show that this does not hold true for all n and they construct a tree T of order $n = 38$ for which $I_d(T) \geq 20 = n/2 + 1$. Moreover, they establish the following result.

Proposition 5.57 [62, Theorem 3] *There exist trees T of arbitrarily large order n such that $I_d(T) \geq \left(\frac{1}{2} + \varepsilon\right) n$ and $I_s(T) \geq \left(\frac{1}{2} + \varepsilon\right) n$ where ε is a small positive constant, independent of T.*

The following general bound on the competition-independence game played on trees with maximum degree at most 3 also holds.

Theorem 5.58 [62, Theorems 4 and 5] *If T is a tree of order $n \geq 2$ and maximum degree at most 3, then*

$$I_d(T) \leq \frac{4}{7}n \quad and \quad I_s(T) \geq \frac{3}{8}n.$$

Goddard and Henning posed the following **independence game $\frac{3}{4}$-Conjecture** for trees.

Conjecture 5.59 [62, Conjecture 1] If T is a tree, then $I_d(T) \leq \frac{3}{4}n(T)$.

As remarked in [62], if Conjecture 5.59 is true, then this conjecture is somewhat sharp, in the sense that there are trees T with $I_d(T) \geq 3n/4 - o(n)$. We close with the following **independence game $\frac{3}{7}$-Conjecture** of Henning.

Conjecture 5.60 [67, Conjecture 14] If T is a tree of order $n \geq 2$, then $I_s(T) \geq \frac{3}{7}n$.

5.7.4 Paired-Domination Game

Given a graph G, a dominating set S with the additional property that the subgraph $G[S]$ induced by S contains a perfect matching (not necessarily induced) is a *paired-dominating set* of G. The *paired-domination number*, $\gamma_{pr}(G)$, of G is the minimum cardinality of a paired-dominating set in G. Haynes and Slater [66] introduced the concept of paired-domination in graphs as a model for assigning backups to guards for security purposes. A survey of paired-domination in graphs can be found in [50].

The paired-domination version of the domination game, which adds a perfect matching dimension to the game, was first investigated by Haynes and Henning [64]. The *paired-domination game*, played on a graph G consists of two players called *Dominator* and *Pairer* who take turns choosing a vertex from G. In this version of the game, each vertex chosen by Dominator must dominate at least one vertex not dominated by the vertices previously chosen, while each vertex chosen by Pairer must be a neighbor of the vertex chosen by Dominator on his previous move that has not previously been chosen. The vertex played by Pairer, together with the vertex played on the previous move by Dominator, are said to be *partners*. A vertex is *unpaired* if it is chosen in a move played by Dominator but does not have a partner. This process eventually produces a paired-dominating set of vertices of G, in which the partners form a matching in the subgraph induced by the set. Dominator wishes to minimize the number of vertices chosen, while Pairer wishes to maximize it. The *game paired-domination number* $\gamma_{gpr}(G)$ of G is the number of vertices chosen when Dominator starts the game and both players play optimally.

In their introductory paper, the authors in [64] prove that Dominator always has a strategy that will finish the game in at most four-fifths the order of the graph with no leaves.

Theorem 5.61 [64, Theorem 6] *If G is a graph with $\delta(G) \geq 2$, then $\gamma_{gpr}(G) \leq \frac{4}{5}n(G)$, and this bound is tight.*

If we impose certain structural restrictions on the graph, then the $\frac{4}{5}$-upper bound on the game paired-domination number given in Theorem 5.61 can be improved to a $\frac{3}{4}$-upper bound.

Theorem 5.62 [64, Theorem 7] *If G is a connected graph with $\delta(G) \geq 2$, then $\gamma_{gpr}(G) \leq \frac{3}{4}n(G)$, if at least one of the following conditions holds:*

(a) *G has no induced subgraph isomorphic to C_4 or C_5;*
(b) *G is 2-connected and bipartite;*
(c) *G is 2-connected and $\deg_G(u) + \deg_G(v) \geq 5$ for every two adjacent vertices u and v.*

Constructions illustrating tightness of the bounds in Theorem 5.62 are given in [64]. We close this subsection with the following two conjectures.

Conjecture 5.63 [64, Conjecture 1] *If G is a bipartite graph with $\delta(G) \geq 2$, then $\gamma_{gpr}(G) \leq \frac{3}{4}n(G)$.*

Conjecture 5.64 [64, Conjecture 2] *If G is a graph with $\delta(G) \geq 3$, then $\gamma_{gpr}(G) \leq \frac{2}{3}n(G)$.*

We remark that if the above conjectures are true, then the bounds are tight as there exists an infinite family of graphs achieving equality in each of these bounds, as constructed in [64].

5.7.5 The Oriented Version of the Domination Game

A *dominating set* of a digraph D with vertex set $V(D)$ and arc set $A(D)$ is a set S of vertices of D such that for every vertex v outside S there exists a vertex $u \in S$ such that there is an arc $(u, v) \in A(D)$ directed from u to v.

In 2002 Alon, Balogh, Bollobás, and Szabó [4] defined the game (oriented) domination number of a graph G a follows. Two players alternately orient an edge of G until all of the edges are oriented. The players have opposite goals, with one player's goal to minimize the domination number of the resulting oriented graph and the other player's to maximize the domination number.

More precisely, the oriented domination game on a graph G consists of two players, *Dominator* and *Avoider*, who take turns to orient the unoriented edges of a graph G, until all edges are oriented. The goal of Dominator is to minimize the domination number of the resulting digraph, while the goal of Avoider is to maximize the domination number. The Dominator-start oriented domination game is the oriented domination game when Dominator plays first. The *game oriented domination number*, $\gamma_{og}(G)$, of G is the minimum possible domination number of the resulting digraph when both players play according to the rules. Alon et al. [4] established the following upper bound on the game oriented domination number of graphs with minimum degree at least 2.

Theorem 5.65 [4, Theorem 4.5] *If G is a graph with $\delta(G) \geq 2$, then*

$$\gamma_{og}(G) \leq \frac{1}{2}n(G).$$

The following conjecture is posed in [4, Conjecture 4.1].

Conjecture 5.66 If G is a graph of order n with maximum degree Δ, then

$$\gamma_{og}(G) \geq \left(\frac{2}{(1 + o(1))\Delta}\right) n.$$

The best general lower bound to date on the game oriented domination number in terms of the maximum degree and order of the graph is the following result.

Theorem 5.67 [4, Theorem 4.2] *If G is a graph of order n with maximum degree Δ, then*

$$\gamma_{og}(G) \geq \left\lceil \frac{4}{3\Delta + 7} \right\rceil n.$$

References

1. Aazami, A.: Hardness results and approximation algorithms for some problems on graphs. ProQuest LLC, Ann Arbor, MI, 2009. Thesis (Ph.D.)–University of Waterloo (Canada)
2. AIM Minimum Rank-Special Graphs Work Group: Zero forcing sets and the minimum rank of graphs. Linear Algebra Appl. **428**, 1628–1648 (2008)
3. Alon, N:. Transversal numbers of uniform hypergraphs. Graphs Combin. **6**, 1–4 (1990)
4. Alon, N., Balogh, J., Bollobás, B., Szabó, T:. Game domination number. Discrete Math. **256**, 23–33 (2002)
5. Bahadır, S., Gözüpek, D., Doğan, O.: k-total uniform graphs (2020). arXiv:2010.08368v2 [math.CO]
6. Bartnicki, T., Brešar, B., Grytczuk, J., Kovše, M., Miechowicz, Z., Peterin, I.: Game chromatic number of Cartesian product graphs. Electron. J. Combin. **15**, Paper 72, 13 pp. (2008)
7. Blank, M.M.: An estimate of the external stability number of a graph without suspended vertices. Prikl. Mat. i Programmirovanie. 3–11, 149 (1973)
8. Borowiecki, M., Fiedorowicz, A., Sidorowicz, E.: Connected domination game. Appl. Anal. Discrete Math. **13**, 261–289 (2019)
9. Brešar, B., Brezovnik, S.: Grundy domination and zero forcing in regular graphs (2020). arXiv:2010.00637 [math.CO]
10. Brešar, B., Bujtás, C., Gologranc, T., Klavžar, S., Košmrlj, G., Marc, T., Patkós, B., Tuza, Z., Vizer, M.: On Grundy total domination number in product graphs. Discuss. Math. Graph Theory. **41**, 225–247 (2021)
11. Brešar, B., Bujtás, C., Gologranc, T., Klavžar, S., Košmrlj, G., Marc, T., Patkós, B., Tuza, Z., Vizer, M.: The variety of domination games. Aequationes Math. **93**, 1085–1109 (2019)
12. Brešar, B., Bujtás, C., Gologranc, T., Klavžar, S., Košmrlj, G., Marc, T., Patkós, B., Tuza, Z., Vizer, M.: Dominating sequences in grid-like and toroidal graphs. Electron. J. Combin. **23**, Paper 4.34, 19 pp. (2016)
13. Brešar, B., Bujtás, C., Gologranc, T., Klavžar, S., Košmrlj, G., Marc, T., Patkós, B., Tuza, Z., Vizer, M.:. Grundy dominating sequences and zero forcing sets. Discrete Optim. **26**, 66–77 (2017)
14. Brešar, B., Dorbec, P., Klavžar, S., Košmrlj, G.: Domination game: effect of edge- and vertex-removal. Discrete Math. **330**, 1–10 (2014)
15. Brešar, B., Dorbec, P., Klavžar, S., Košmrlj, G.: How long can one bluff in the domination game? Discuss. Math. Graph Theory **37**, 337–352 (2017)
16. Brešar, B., Dorbec, P., Klavžar, S., Košmrlj, G., Renault, G.: Complexity of the game domination problem. Theoret. Comput. Sci. **648**, 1–7 (2016)

17. Brešar, B., Gologranc, T., Henning, M., Kos, T.: On the L-Grundy domination number of a graph. Filomat **34**, 3205–3215 (2020)
18. Brešar, B., Gologranc, T., Kos, T.: Dominating sequences under atomic changes with applications in Sierpiński and interval graphs. Appl. Anal. Discrete Math. **10**, 518–531 (2016)
19. Brešar, B., Gologranc, T., Milanič, M., Rall, D.F., Rizzi, R.: Dominating sequences in graphs. Discrete Math. **336**, 22–36 (2014)
20. Brešar, B., Henning, M.A.: The game total domination problem is log-complete in PSPACE. Inform. Process. Lett. **126**, 12–17 (2017)
21. Brešar, B., Henning, M.A., Rall, D.F.: Total dominating sequences in graphs. Discrete Math. **339**, 1665–1676 (2016)
22. Brešar, B., Klavžar, S., Košmrlj, G., Rall, D.F.: Domination game: extremal families of graphs for 3/5-conjectures. Discrete Appl. Math. **161**, 1308–1316 (2013)
23. Brešar, B., Klavžar, S., Košmrlj, G., Rall, D.F.: Guarded subgraphs and the domination game. Discrete Math. Theor. Comput. Sci. **17**, 161–168 (2015)
24. Brešar, B., Klavžar, S., Rall, D.F.: Domination game and an imagination strategy. SIAM J. Discrete Math. **24**, 979–991 (2010)
25. Brešar, B., Klavžar, S., Rall, D.F.: Domination game played on trees and spanning subgraphs. Discrete Math. **313**, 915–923 (2013)
26. Brešar, B., Kos, T., Nasini, G., Torres, P.: Total dominating sequences in trees, split graphs, and under modular decomposition. Discrete Optim. **28**, 16–30 (2018)
27. Brešar, B., Kos, T., Nasini, G., Torres, P.: Grundy domination and zero forcing in Kneser graphs. Ars Math. Contemp. **17**, 419–430 (2019)
28. Bujtás, C.: Domination number of graphs with minimum degree five. Discuss. Math. Graph Theory **41**, 763–777 (2021)
29. Bujtás, C.: Domination game on forests. Discrete Math. **338**, 2220–2228 (2015)
30. Bujtás, C.: On the game domination number of graphs with given minimum degree. Electron. J. Combin. **22**, Paper 3.29, 18 pp. (2015)
31. Bujtás, C.: On the game total domination number. Graphs Combin. **34**, 415–425 (2018)
32. Bujtás, C.: General upper bound on the game domination number. Discrete Appl. Math. **285**, 530–538 (2020)
33. Bujtás, C., Dokyeesun, P., Iršič, V., Klavžar, S.: Connected domination game played on Cartesian products. Open Math. **17**, 1269–1280 (2019)
34. Bujtás, C., Henning, M.A., Tuza, Z.: Transversal game on hypergraphs and the $\frac{3}{4}$-conjecture on the total domination game. SIAM J. Discrete Math. **30**, 1830–1847 (2016)
35. Bujtás, C., Henning, M.A., Tuza, Z.: Bounds on the game transversal number in hypergraphs. Eur. J. Combin. **59**, 34–50 (2017)
36. Bujtás, C., Iršič, V., Klavžar, S.: 1/2-conjectures on the domination game and claw-free graphs (2020). arXiv:2010.14273 [math.CO]
37. Bujtás, C., Iršič, V., Klavžar, S.: Perfect graphs for domination games Ann. Comb. **25**, 133–152 (2021)
38. Bujtás, C., Iršič, V., Klavžar, S.: Z-domination game. Discrete Math. **343**, 112076 (2020)
39. Bujtás, C., Iršič, V., Klavžar, S., Xu, K.: The domination game played on diameter 2 graphs Aequat. Math. (2021). https://doi.org/10.1007/s00010-021-00786-x
40. Bujtás, C., Iršič, V., Klavžar, S., Xu, K.: On Rall's 1/2-conjecture on the domination game. Quaest. Math. (2020). https://doi.org/10.2989/16073606.2020.1822945
41. Bujtás, C., Jaskó, S.: Bounds on the 2-domination number. Discrete Appl. Math. **242**, 4–15 (2018)
42. Bujtás, C., Klavžar, S.: Improved upper bounds on the domination number of graphs with minimum degree at least five. Graphs Combin. **32**, 511–519 (2016)
43. Bujtás, C., Klavžar, S., Košmrlj, G.: Domination game critical graphs. Discuss. Math. Graph Theory **35**, 781–796 (2015)
44. Bujtás, C., Patkós, B., Tuza, Z., Vizer, M.: Domination game on uniform hypergraphs. Discrete Appl. Math. **258**, 65–75 (2019)
45. Bujtás, C., Tuza, Z.: The disjoint domination game. Discrete Math. **339**, 1985–1992 (2016)

46. Bujtás, C., Tuza, Z.: Fractional domination game. Electron. J. Combin. **26**, Paper 4.3 (2019)
47. Charoensitthichai, K., Worawannotai, C.: Total domination game on ladder graphs. Songklanakarin J. Sci. Technol. (in press)
48. Charoensitthichai, K., Worawannotai, C.: Effect of vertex-removal on game total domination numbers. Asian-Eur. J. Math. **13**(7), 2050129 (2020)
49. Chvátal, V., McDiarmid, C.J.H.: Small transversals in hypergraphs. Combinatorica **12**, 19–26 (1992)
50. Desormeaux, W.J., Henning, M.A.: Paired domination in graphs: a survey and recent results. Util. Math. **94**, 101–166 (2014)
51. Dorbec, P., Henning, M.A.: Game total domination for cycles and paths. Discrete Appl. Math. **208**, 7–18 (2016)
52. Dorbec, P., Henning, M.A., Klavžar, S., Košmrlj, G.: Cutting lemma and union lemma for the domination game. Discrete Math. **342**, 1213–1222 (2019)
53. Dorbec, P., Košmrlj, G., Renault, G.: The domination game played on unions of graphs. Discrete Math. **338**, 71–79 (2015)
54. Dravec, T., Jakovac, M., Kós, T., Marc, T.: On graphs with equal total domination and Grundy total domination numbers. Aequationes Math. (2021). https://doi.org/10.1007/s00010-021-00776-z
55. Duchêne, É., Gledel, V., Parreau, A., Renault, G.: Maker-Breaker domination game. Discrete Math. **343**, Paper 111955, 12 pp. (2020)
56. Erdős, P., Selfridge, J.L.: On a combinatorial game. J. Combin. Theory Ser. A **14**, 298–301 (1973)
57. Erdős, P., Tuza, Z.: Vertex coverings of the edge set in a connected graph. In: Graph Theory, Combinatorics, and Algorithms, Vol. 1, 2 (Kalamazoo, MI, 1992), Wiley-Intersci. Publ., pp. 1179–1187. Wiley, New York (1995)
58. Erey, A.: Length uniformity in legal dominating sequences. Graphs Combin. **36**, 1819–1825 (2020)
59. Fallat, S.M., Hogben, L.: The minimum rank of symmetric matrices described by a graph: a survey. Linear Algebra Appl. **426**, 558–582 (2007)
60. Gledel, V., Henning, M.A., Iršič, V., Klavžar, S.: Maker-Breaker total domination game. Discrete Appl. Math. **282**, 96–107 (2020)
61. Gledel, V., Iršič, V., Klavžar, S.: Maker-Breaker domination number. Bull. Malays. Math. Sci. Soc. **42**, 1773–1789 (2019)
62. Goddard, W., Henning, M.A.: The competition-independence game in trees. J. Combin. Math. Combin. Comput. **104**, 161–170 (2018)
63. Hammack, R., Imrich, W., Klavžar, S.: Handbook of product graphs. Discrete Mathematics and its Applications (Boca Raton), 2nd edn. CRC Press, Boca Raton (2011). With a foreword by Peter Winkler
64. Haynes, T., Henning, M.A.: Paired-domination game played in graphs. Commun. Combin. Optim. **4**, 79–94 (2019)
65. Haynes, T.W., Hedetniemi, S.T., Slater, P.J.: Fundamentals of Domination in Graphs. Monographs and Textbooks in Pure and Applied Mathematics, vol. 208. Marcel Dekker, New York (1998)
66. Haynes, T.W., Slater, P.J.: Paired-domination in graphs. Networks **32**, 199–206 (1998)
67. Henning, M.A.: My favorite domination game conjectures. In: Graph Theory—Favorite Conjectures and Open Problems, vol. 2. Probl. Books in Math., pp. 135–148. Springer, Cham (2018)
68. Henning, M.A., Kinnersley, W.B.: Domination game: a proof of the 3/5-conjecture for graphs with minimum degree at least two. SIAM J. Discrete Math. **30**, 20–35 (2016)
69. Henning, M.A., Klavžar, S.: Infinite families of circular and Möbius ladders that are total domination game critical. Bull. Malays. Math. Sci. Soc. **41**, 2141–2149 (2018)
70. Henning, M.A., Klavžar, S., Rall, D.F.: Total version of the domination game. Graphs Combin. **31**, 1453–1462 (2015)

71. Henning, M.A., Klavžar, S., Rall, D.F.: The 4/5 upper bound on the game total domination number. Combinatorica **37**, 223–251 (2017)

72. Henning, M.A., Klavžar, S., Rall, D.F.: Game total domination critical graphs. Discrete Appl. Math. **250**, 28–37 (2018)

73. Henning, M.A., Löwenstein, C.: Domination game: extremal families for the 3/5-conjecture for forests. Discuss. Math. Graph Theory **37**, 369–381 (2017)

74. Henning, M.A., Rall, D.F.: The enclaveless competition game. Ars Math. Contemp. (2020). https://doi.org/10.26493/1855-3974.2227.e1a

75. Henning, M.A., Rall, D.F.: Progress towards the total domination game $\frac{3}{4}$-conjecture. Discrete Math. **339**, 2620–2627 (2016)

76. Henning, M.A., Rall, D.F.: Trees with equal total domination and game total domination numbers. Discrete Appl. Math. **226**, 58–70 (2017)

77. Henning, M.A., Yeo, A.: Total Domination in Graphs. Springer Monographs in Mathematics. Springer, New York (2013)

78. Hogben, L.: Minimum rank problems. Linear Algebra Appl. **432**, 1961–1974 (2010)

79. Iršič, V.: Connected domination game: predomination, Staller-start game, and lexicographic products (2019). arXiv:1902.02087v2 [math.CO]

80. Iršič, V.: Effect of predomination and vertex removal on the game total domination number of a graph. Discrete Appl. Math. **257**, 216–225 (2019)

81. James, T., Dorbec, P., Vijayakumar, A.: Further progress on the heredity of the game domination number. In: Theoretical Computer Science and Discrete Mathematics. Lecture Notes in Comput. Sci., vol. 10398, pp. 435–444. Springer, Cham (2017)

82. James, T., Klavžar, S., Vijayakumar, A.: The domination game on split graphs. Bull. Aust. Math. Soc. **99**, 327–337 (2019)

83. James, T., Vijayakumar, A.: Domination game: effect of edge contraction and edge subdivision. Discuss. Math. Graph Theory. (in press)

84. Jayaram, B., Arumugam, S., Thulasiraman, K.: Dominator sequences in bipartite graphs. Theoret. Comput. Sci. **694**, 34–41 (2017)

85. Jiang, Y., Lu, M.: Game total domination for cyclic bipartite graphs. Discrete Appl. Math. **265**, 120–127 (2019)

86. Kinnersley, W.B., West, D.B., Zamani, R.: Extremal problems for game domination number. SIAM J. Discrete Math. **27**, 2090–2107 (2013)

87. Klavžar, S., Košmrlj, G., Schmidt, S.: On the computational complexity of the domination game. Iran. J. Math. Sci. Inform. **10**, 115–122 (2015)

88. Klavžar, S., Košmrlj, G., Schmidt, S.: On graphs with small game domination number. Appl. Anal. Discrete Math. **10**, 30–45 (2016)

89. Klavžar, S., Rall, D.F.: Domination game and minimal edge cuts. Discrete Math. **342**, 951–958 (2019)

90. Kos, T.: Contributions to the study of contemporary domination invariants of graphs. University of Maribor, Doctoral Dissertation, (2019)

91. Kostochka, A.V., Stodolsky, B.Y.: An upper bound on the domination number of n-vertex connected cubic graphs. Discrete Math. **309**, 1142–1162 (2009)

92. Košmrlj, G.: Realizations of the game domination number. J. Comb. Optim. **28**, 447–461 (2014)

93. Košmrlj, G.: Domination game on paths and cycles. Ars Math. Contemp. **13**, 125–136 (2017)

94. Lin, J.C.-H.: Zero forcing number, Grundy domination number, and their variants. Linear Algebra Appl. **563**, 240–254 (2019)

95. Marcus, N., Peleg, D.: The domination game: Proving the 3/5 conjecture on isolate-free forests (2016). arXiv:1603.01181 [cs.DS]

96. McCuaig, W., Shepherd, B.: Domination in graphs with minimum degree two. J. Graph Theory **13**, 749–762 (1989)

97. Nadjafi-Arani, M.J., Siggers, M., Soltani, H.: Characterisation of forests with trivial game domination numbers. J. Comb. Optim. **32**, 800–811 (2016)

98. Nasini, G., Torres, P.: Grundy dominating sequences on X-join product. Discrete Appl. Math. **284**, 138–149 (2020)

99. Ore, O.: Theory of Graphs. American Mathematical Society Colloquium Publications, vol. XXXVIII. American Mathematical Society, Providence (1962)

100. Phillips, J.B., Slater, P.J.: An introduction to graph competition independence and enclaveless parameters. Graph Theory Notes N.Y. **41**, 37–41 (2001)

101. Phillips, J.B., Slater, P.J.: Graph competition independence and enclaveless parameters. In: Proceedings of the Thirty-third Southeastern International Conference on Combinatorics, Graph Theory and Computing (Boca Raton, FL, 2002), vol. 154, pp. 79–100 (2002)

102. Reed, B.: Paths, stars and the number three. Combin. Probab. Comput. **5**, 277–295 (1996)

103. Row, D.D.: A technique for computing the zero forcing number of a graph with a cut-vertex. Linear Algebra Appl. **436**, 4423–4432 (2012)

104. Ruksasakchai, W., Onphaeng, K., Worawannotai, C.: Game domination numbers of a disjoint union of paths and cycles. Quaest. Math. **42**, 1357–1372 (2019)

105. Schaefer, T.J.: On the complexity of some two-person perfect-information games. J. Comput. System Sci. **16**, 185–225 (1978)

106. Schmidt, S.: The 3/5-conjecture for weakly $S(K_{1,3})$-free forests. Discrete Math. **339**, 2767–2774 (2016)

107. Sohn, M.Y., Xudong, Y.: Domination in graphs of minimum degree four. J. Korean Math. Soc. **46**, 759–773 (2009)

108. Tuza, Z.: Covering all cliques of a graph. Discrete Math. **86**, 117–126 (1990)

109. West, D.B.: Introduction to Graph Theory. Prentice Hall, Upper Saddle River (1996)

110. Worawannotai, C., Chantarachada, N.: Game domination numbers of a disjoint union of chains and cycles of complete graphs. Chamchuri J. Math. **11**, 10–25 (2019)

111. Xu, K., Li, X.: On domination game stable graphs and domination game edge-critical graphs. Discrete Appl. Math. **250**, 47–56 (2018)

112. Xu, K., Li, X., Klavžar, S.: On graphs with largest possible game domination number. Discrete Math. **341**, 1768–1777 (2018)

113. Zhu, X.: Game coloring the Cartesian product of graphs. J. Graph Theory **59**, 261–278 (2008)

Author Index

Subject Index

© The Author(s), under exclusive license to Springer Nature Switzerland AG 2021
B. Brešar et al., *Domination Games Played on Graphs*, SpringerBriefs in
Mathematics, https://doi.org/10.1007/978-3-030-69087-8

Symbol Index

© The Author(s), under exclusive license to Springer Nature Switzerland AG 2021

B. Brešar et al., *Domination Games Played on Graphs*, SpringerBriefs in Mathematics, https://doi.org/10.1007/978-3-030-69087-8

Printed in the United States
by Baker & Taylor Publisher Services